General Preface to the Series

Because it is no longer possible for one textbook to cover the whole field of biology while remaining sufficiently up to date, the Institute of Biology has sponsored this series so that teachers and students can learn about significant developments. The enthusiastic acceptance of 'Studies in Biology' shows that the books are providing authoritative views of biological topics.

The features of the series include the attention given to methods, the selected list of books for further reading and, wherever possible, suggestions for practical work.

Readers' comments will be welcomed by the Education Officer of the Institute.

1979   Institute of Biology
                                              41 Queen's Gate
                                              London SW7 5BR

Preface

In the years that have elapsed since the first publication of this book a wealth of new information about animal behaviour has accumulated, and some of this has led to changes in concepts. In a book of this size it is possible to make only passing reference to many important issues, and some have been neglected altogether, but I have tried to give more emphasis to the development of behaviour, to communication and defensive behaviour, and to the importance of these three aspects in social behaviour. This has meant the replacement of one chapter and re-writing of others to a greater or lesser extent, in which I have tried to conserve the essential idiom and character of the book as it was written by the late John Carthy. If I have succeeded to any extent in this, it is my small tribute to a man of great charm who I was very fortunate to have known.

I am grateful to Drs Donald Broom, Rory Putman and Peter Lockwood for useful criticisms they have made of the typescript, and to Dorothy Carthy for her encouragement.

1978   P. E. H.

Contents

The Institute of Biology's
Studies in Biology no. 3

The Study of
Behaviour

Second Edition

J. D. Carthy
revised by

E. Howse
nior lecturer in Biology
ersity of Southampton

nold

© Philip E. Howse, 1979

First published 1966
by Edward Arnold (Publishers) Limited
41 Bedford Square, London WC1B 3DQ

Reprinted 1967
Reprinted 1970
Reprinted 1971
Reprinted 1973
Reprinted 1975
Second Edition 1979

British Library Cataloguing in Publication Data

Carthy, John Dennis
 The study of behaviour.   2nd ed. – (Institute of
 Biology.   Studies in biology; no. 3   ISSN 0537-9024).
 1.   Animals, Habits and behavior of
 I.   Title   II.   Howse, Philip Edwin   III.   Series
 591.5   QL751
 ISBN 0-7131-2749-X

Printed and bound in Great Britain at
The Camelot Press Ltd, Southampton

1 Observing and Describing Behaviour

1.1 What is behaviour?

What we call behaviour is what we perceive of an animal's reactions to the environment around it which are, in their turn, influenced by variable internal factors. These reactions usually involve movement. A blowfly larva moves away from light, for example, and birds raise their wings or erect their feathers in displays which form part of their behavioural repertoire. Sometimes the reactions may be less obvious. A cuttlefish about to close in on its prey will flush with bands of colour which travel along its body. These colour changes are produced by the co-ordinated activity of several kinds of chromatophores in the skin, which is no less a behavioural response than are the courtship movements of the same cuttlefish. The chromatophores are effectors, like the muscles that move the limbs.

The study of behaviour begins with observations of an animal's movements, posture, and so forth. Very often an animal may appear to be doing nothing, even if its surroundings change. This may be because it fails to perceive the changes, but it is equally possible that its response to the changes may be to remain still. If a guppy is introduced into a new tank it will usually dart to the bottom and 'freeze', remaining quite still even in the face of large disturbances. This is part of the behaviour which leads to avoidance of predators. A tortoise-shell butterfly may rest for long periods on a wall in the sun with its wings extended. It is actually gaining heat in the process, and the position of the wings relative to the sun's rays is very important and may vary subtly as the body heats up. Usually, only by close and repeated observations of animals in natural or near natural surroundings is it possible to recognize behaviour like this and begin to see how it is related to the environmental stimuli.

There are five important questions that we can ask about behaviour. What causes it? What is its function? How does the behaviour pattern develop? How did it evolve? How much can it be changed during the lifetime of the individual? Many of these are questions which can also be asked about a morphological feature and there are many parallels between the evolution of structural features and behavioural ones. Many behavioural patterns, like morphological features, are very resistant to change by environmental factors (i.e. by what the animal experiences) but it is important to note that the experience of the animal can result in radical changes in some kinds of behaviour in certain circumstances – a process we know as learning. All behaviour, though, depends upon environmental factors of one kind or another.

1.2   Recording behaviour

The first aim of the study of behaviour of a particular animal is to record it in detail, correlating it with stimuli that evoke the different sections of it. Such a complete catalogue of behaviour is called an ethogram. It is vitally important that such an ethogram should be recorded quite impartially. The observer must not be influenced by his own evaluation of what is going on but must seek to record everything, however unimportant it may seem at the time. Even details of weather conditions may later prove necessary in analysing the causation of the behaviour. Care must be taken to avoid anthropomorphism (imputing human motives to the animal), although this is often a useful form of shorthand. Animal behaviour was only able to advance as an objective science when terms like 'desire', 'purpose', 'decide' and so forth, which give explanations in terms of human mental activity, were left out of consideration. The report must therefore be a plain recording on the basis of which simple hypotheses can be constructed, which do not view the animal as a complex thinking being, and put to the test experimentally.

It is rarely possible to describe a behaviour pattern after a single observation. It is necessary to know the range of circumstances in which the behaviour occurs and also the range of variations of that behaviour which can be performed by the animal. It may be some time before the observations begin to fall into an ordered whole which reveals something of the causation of the behaviour, its function and relationship to other behaviour patterns in the animal's repertoire. However, experience increases the ease with which subtle changes in behaviour can be recognized and those who have earned the highest reputation for their work have generally been the most patient observers. The great French entomologist, Fabre, often sat in the same spot observing insect behaviour from dawn to dusk and passers-by thought him demented.

A number of techniques can be used to augment simple visual observation, some of them very expensive and elaborate, others very simple. Sketches of behaviour, even if they are very diagrammatic and rudimentary, enable a start to be made in categorizing behaviour. It is usually most important to be able to recognize individual animals from each other. Often small variations in size or colour patterns can be used, but marking may be necessary. If so, there are many methods. Coloured leg-rings can be put on birds (though this can be done only by licensed people). Cellulose paint adheres well to most insects, particularly if their bodies are hairy. If the thorax is marked with spots of three different colours in different arrangements a large number of insects can be separated. This method has been used for honeybees by Ribbands, who used not spots but certain letter-like symbols which could still be distinguished if parts were cleaned or rubbed off. Fish can be marked by removing certain scales or 'freeze branded' with a very cold metal rod, and mammals by clipping off fur or toe-nails, but these are procedures that can only be done in Britain under a government licence.

Photography is a great help in analysing behaviour because the camera provides an impartial record. But even this has limitations because a camera records only what is going on in front of it. The noise of a cine-camera can be disturbing to animals, and the proximity of a camera may also be intrusive. Eibl-Eibesfeldt has been able to study natural human facial expressions only by pointing a camera away from the subject and taking pictures surreptitiously with a lens set in at the side.

A written description of behaviour can be done in shorthand or in note form. When it is a question of scoring the occurrence of a behaviour pattern a simple *pro forma* can be devised which the observer fills in with marks in the appropriate column as the behaviour is going on. Time is an important dimension in behaviour and it is not always easy to keep looking at a watch while recording observations. This can be overcome by putting records or marks on a paper-covered drum (such as a kymograph drum) revolving at a known speed. Cassette tape recorders fulfil the same function and are easy to use in the field. Cinematograph film and taped recordings can be played over and over again, so that details can be picked out. The same is true of video-tape recordings, which have been used more and more in recent years, especially for studying the behaviour of animals under captive conditions. A closed-circuit television camera can be mounted permanently and unobtrusively in a cage or enclosure and does not require an operator near it so the animals are submitted to a minimum of disturbance.

Cassette recorders have a limited usefulness for recording many animal sounds because these instruments usually have a very restricted frequency range and distort the frequencies they record. For more accurate work recorders should be chosen which have a range extending up to and around 18 kHz and have a microphone with a corresponding sensitivity. Directional microphones help to cut down background noise, and directionality can be improved in the field if a parabolic reflector (or even a home-made cone) is used. Good recordings can be analysed on a sound spectrograph which plots the frequencies present in the sound against time (Fig. 1–1). Adequate notes must be taken of the circumstances under which the sounds were recorded. Recordings can be played back to animals to see what their responses are and thus to determine the significance of the sounds for them.

1.3   Studying behaviour experimentally

In the 1930s there was an intense interest in the behaviour of invertebrates such as crabs and robber flies in beams of light or on sloping surfaces in controlled laboratory conditions. The complex behaviour of the animals in the field, during courtship and so forth, was overlooked and probably unknown to most of the experimenters. It is important to /

know the natural behaviour of the animal in its normal environment before analysing particular parts of it in the laboratory otherwise the wrong conclusions can easily be drawn. Carefully controlled laboratory conditions may be so different from the animal's normal environment that it may be inhibited from showing many kinds of behaviour and may even do nothing at all. Cold-blooded animals are dependent upon temperature and show certain types of activity like courtship or feeding only if their body temperature falls within a certain range. Such behaviour in most animals is also linked to the time of day. Some butterflies are active in field only during part of the morning, others during the afternoon. Crickets start to sing most intensively at dusk, and mice are far more active at night. It is inconvenient to stay up all night to watch the behaviour of laboratory mice, but with the aid of a time switch their lighting cycles and activity cycles can easily be reversed and they can be observed during daylight hours in dim light.

It is important also to give an animal time to become familiar with its environment unless one is actually investigating its response to a change in conditions. If, for example, fish are introduced into a new tank their behaviour tends to become more varied and their attitudes less fearful over a period of days. Many animals, including rats and mice, treat their cage as a territory after they have explored it and behave aggressively towards newcomers.

Controls must be used in experiments on behaviour, or control situations examined, just as they are in other experimental approaches. In the study of behaviour, science is measurement, as it is in other disciplines. Finding the right thing to measure is not always easy. If an animal repeats an activity a number of times and then switches to doing something else for a time, the activities can be grouped as 'bouts' and their duration and the intervals between them measured. If the behaviour that is being examined recurs frequently it need not be recorded all the time, but only during (say) five-minute periods at hourly intervals. Another method is to record what the animal is doing at regular intervals (say every ten minutes). Sampling methods such as these can greatly simplify experiments, but behavioural data commonly show a great deal of variability, and it is important in drawing graphs or histograms of results to indicate the range of this variability in some way.

Simply by observing behaviour it is not possible to decide which are the stimuli that evoke it. For one thing, it is not possible to see all the stimuli that may be responsible; for another, the sensory capacities of animals can be very different from our own. Honeybees can see light in the ultraviolet region and detect plane polarized light, pigeons can detect weak magnetic fields, some fish can detect electric fields, many animals can hear ultrasound, and some insects respond strongly to chemicals that are odourless to us. There are three main ways in which the sensory abilities of animals can be explored. By measuring the responses of an animal to a range of simple models, starting with one that readily elicits a response of some kind and then changing the colour, shape, or whatever is under investigation. By training animals to respond to a particular stimulus by, for example, giving them food when they do so, and then seeing how far they can discriminate that stimulus from a range of similar ones when given a choice. And, more objectively, by using electrodes to record the nerve impulses from the sense organ in question.

The sense organ may itself select certain aspects of the stimuli and not pass on all the information contained in them to the central nervous system (CNS). In this way it functions as a filter and makes the task of the CNS easier (Chapter 3). For example, the female silkmoth, *Bombyx mori* produces an attractant for males, known as bombykol. Sensilla on the antennae of the males are extremely sensitive to bombykol and a few chemicals that are very similar in structure, but do not respond to other odours. Nerve impulses travelling along the nerves from these sensilla thus have only one meaning to the CNS. The sense organs of mammals, on the other hand, do relatively little filtering, although they do do some, and much is left to the brain, necessarily a much more complex structure than in insects, to evaluate. Among the influences which affect the nature of efferent signals controlling the response of the animal, and which must be taken into account in evaluating experimental data, are its genetic make-up, its present physiological state (for example, whether its gonads are actively producing hormones) and its acquired experience. So there is rarely a direct relationship between stimulus and response; there are many factors operating between the observed beginning and ending of the behavioural process.

1.4   Behavioural concepts

The founders of the so-called ethological school of animal behaviour, Lorenz and Tinbergen, helped to develop an extensive terminology of behaviour, in which they separated the concepts of reflexes, orientation behaviour, instinct and learning. Reflexes are commonly thought of as simple automatic responses involving only part of the nervous system and not the brain. These were studied in detail by the physiologist Sir Charles Sherrington who suggested that the simplest element of behaviour was an arc including a sensory nerve, perhaps only one internuncial neuron, and a motor neuron. Such simple reflexes may, in fact, not exist except in some simple invertebrates, and reflexes like swallowing or breathing involve the co-ordinated activity of a great many nerves and muscles. Reflexes differ from other forms of behaviour probably only because they are without the modulating activity of the brain. Lorenz separated reflexes and instinct by saying that an instinct has a central inflexible core (the fixed action pattern) which generates a kind of nervous energy which influences the threshold for the behaviour. He separated instinct and learning by maintaining that instincts were innate, and not dependent upon experience, while learning was the converse – behaviour that is entirely acquired as a result of the individual's experience. Now instinctive patterns of behaviour cannot be entirely inflexible – even if an animal is performing a stereotyped display it needs to make continual postural adjustments and to remain facing in the right direction. It was therefore suggested that the fixed action pattern had a 'coat' of reflexes and orientation movements.

These concepts are nowadays seen to be quite artificial distinctions. Mistakes have been made by assuming that stereotyped species-typical behaviour is instinctive and therefore is innate. The songs of chaffinches could be described in this way, but Thorpe showed that young chaffinches can sing only a rudimentary song (Fig. 1–1). They modify this as a result of

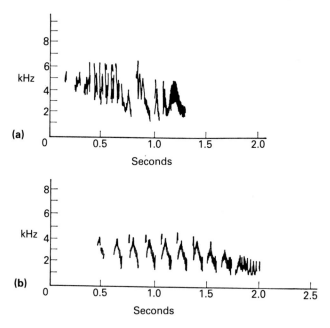

Fig. 1–1   Sound spectrograms of (a) a normal chaffinch song and (b) a song from a bird reared in isolation. These graphs, produced electronically, represent the frequencies of the sound (in kHz) during a phase of song lasting 1–2 s. (After THORPE, W. H. (1958). *Ibis*, **100**, 545.)

listening to their parents near the end of their first season and thereafter it cannot be changed. Even birds which are allowed to hear only their own voice sing slightly better songs than those which are deafened. Considerable variations in song pattern occur in the shrike, *Laniarius aethiopicus*, in which the male and female of a pair sing duets, and the Indian hill mynah and many members of the parrot family are capable of considerable feats of vocal imitation. At the other end of the scale, the songs of American juncos appear not to contain any acquired elements.

So, as far as bird song goes, basic patterns are more or less susceptible to change as a result of various kinds of environmental influence, depending on the species. To call a song instinctive or learnt without experimental evidence would pre-judge the issue and would probably not be correct in any circumstances, because behaviour, even more so than any structural feature, cannot develop apart from the environment and without feed-back from it. What applies to bird song must apply to any behaviour pattern.

2 Orientation and Navigation

2.1 Orientation

Locomotion is often, but not always, directed, for an animal usually moves in a way which is orientated with regard to a source of stimulation or which results in the animals being grouped with respect to a stimulus source. The distribution of animals in an environment is influenced by their behavioural reactions to various physical and chemical factors in that environment and often also to each other. Responses to humidity may determine whether they remain in damp places or come out into the open; responses to light may determine where they will rest and so on.

The study of the orientation reactions of animals to stimuli provided the material for a classification of the kinds of reaction which are observed under laboratory conditions where single stimuli of controlled intensity and direction can be used. Though such a classification may be a convenience the terms used should not be taken as explaining how the behaviour comes about. Too often the names used to describe the behavioural reactions have been taken to be the names of the mechanisms determining those reactions.

First, there are the kinds of behaviour in which the animal's movements are not orientated with respect to the direction of the stimulus source. These are the *kineses*. They can be shown very well in experiments with choice chambers. These chambers can be constructed in many ways but they are essentially a round arena floored usually with gauze and divided in two along a diameter (Fig. 2–1). One half may be shaded with black paper and the whole illuminated with overhead light for experiments on light reactions. Beneath the floor may be placed potassium hydroxide or sulphuric acid solutions to give particular humidities in the arena above. Similarly various odour sources may be put beneath the floor. In one form of kinesis the animal's speed of movement is affected by whether or not it is being stimulated. Thus, woodlice placed in a choice chamber, one half of which has air of high humidity in it and the other low humidity, will, after a time, collect in the half with high humidity. This in itself is one observation, the final distribution of the animals shows them to be positive in their reaction to humidity. But of interest as well is the minute by minute response of an individual. Each one runs about in the chamber at first, but runs much faster in the dry side than in the wet. Indeed they move so slowly in the wet that they sometimes even come to a stop. So another observation may be the speed of their movement in the two halves obtained by following the tracks of individuals. Again the effect of various humidities can be shown by putting, say, twenty woodlice in a

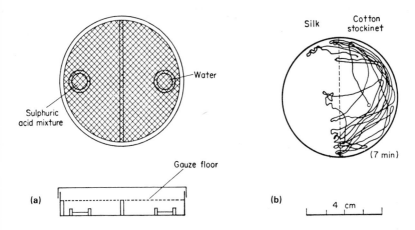

Silk

Cotton stockinet

Water

Sulphuric acid mixture

Gauze floor

(a)

(b)

(7 min)

4 cm

Fig. 2–1 (a) A choice chamber in section and elevation. This one is arranged to have a saturated atmosphere on one side and one of a lower humidity on the other. (b) Klinokinetic behaviour of lice on different materials. The louse turns violently when it moves off the stockinet onto the smooth silk. ((b) After WIGGLESWORTH, V. B. (1941). *Parasitology*, 33, 67.)

chamber which is maintained at a particular humidity. The percentage of them which are active at each humidity can be recorded. All these experiments show that the distribution of the woodlice can be explained in some circumstances by a change of speed inversely correlated with relative humidity imposed upon randomly directed locomotion. This is *orthokinesis*. The animals cluster in the area of optimal conditions merely because they move more slowly there and therefore must spend more time there.

The term *klinokinesis* has been used a great deal to describe the orientation of animals such as flatworms which were said to increase their rate of turning in high light intensity and then to orientate down a light gradient, finishing up in the dark. The work on which this was based has been found to be unsound, and so klinokinesis is now mostly used to describe the behaviour of animals approaching a source of a favourable stimulus. If the animal continues in a straight path, it will eventually begin to move out of the favourable zone, and then start to make turning movements which will ensure that it orientates up the stimulus gradient. Thus an increase in the intensity of stimulation gives rise to an increase in the rate of turning, although the direction of the turning movements themselves may be random.

In orthokinesis and klinokinesis, animals do not maintain a fixed direction of movement, but in the other kinds of orientation, the *taxes*, their paths are always at a fixed angle to the stimulus source, or may lead directly towards or away from the source. Since most of the work on these behaviours has been done using light stimuli, examples will mainly be taken from that type of stimulation. In addition, stimulus situations with light sources are relatively easy to set up.

Such a taxis movement can be brought about in a number of ways. One is illustrated by the movements of a blowfly larva when it is put on a horizontal surface in a dark room and illuminated by a torch resting on the surface. If the larva is one which is about to pupate, it moves away from the torch bulb down the beam. If another torch pointing across its path is next switched on, in place of the first one, it then turns away from the second source. As the larva moves it swings its head from side to side. It has two symmetrically placed light receptors on its head. When it swings to the left the left-hand one is exposed to the light behind it; this evokes a bend to the right, away from the light. But on swinging to the right, the right-hand receptor is illuminated, which causes a swing to the left. These alternately evoked swings make the animal move in a roughly straight line. This directed movement involving alternate testing of the environment by bending its body is called a *klinotaxis*. The role of the head movement in testing the environment is confirmed when a larva is allowed to crawl in a dish in red light and is illuminated by an overhead light only when it swings to the right. The larva will then move in a circle to the left, because left turns will be produced by the illumination on the right swing only.

In *tropotaxis*, animals orientate without bending from side to side as they move. Such a movement is supposed to depend upon symmetrically placed receptors the stimulation of which is compared simultaneously in the central nervous system. In a klinotactic behaviour, the stimulation of the receptors must be compared successively and thus involves a very short-term 'memory'. If, for example, a hoverfly has one eye covered with black cellulose paint, and is placed in an arena which is evenly lit from above, it will walk in circles always turning to the side of its seeing eye. An animal which is negatively phototactic will turn towards its blinded side. The blinding has prevented the normal balance of stimulation and causes the circling movements.

The third sub-division of this classification contains many kinds of orientation behaviour. Essentially it consists of behaviour which is called *telotaxis*. The path of an animal behaving in a telotactic manner is orientated whether or not both symmetrical sense organs are functional, one being sufficient. A honeybee will walk up a beam of light towards a light source (clipping the bee's wings will prevent it flying) even if one eye is painted over. A hoverfly will do the same. This behaviour can be explained if the stimulation of the ommatidia pointing to the side and back of, say, the right eye evokes strong turning of the whole body to the right, so that the head is swung round towards the light, and if stimulation at points closer and closer to the front of the animal causes less and less strong turning. Clearly more violent turning will be necessary to bring the animal to face a light source behind it than one nearly ahead of it. There is a small area of the eye pointing ahead, a light on which evokes no turning at all. Within this area and between it and the mid-line of the head there are ommatidia which in the right eye cause turning to the left. As a guidance mechanism such an arrangement can produce accurate orientation elements like these even in one eye alone.

There is a test which can be used to distinguish further between the three types of taxis. This is the two-light test. If two lights of equal intensity are placed in front of a photo-positive animal, it may move between the lights or directly to one or other of them. A klinotactic or a tropotactic animal will move between the lights on a path lying at an equal distance from both. If one light is stronger than the other, the path will incline towards the weaker light. Such paths will, of course, be those on which the animals will be equally stimulated by both sources. A telotactic animal, on the other hand, heads straight to one or other of the lamps. Such behaviour must involve selection, as the animal will inevitably be stimulated by both lights and must ignore one in moving towards the other and may even switch attention from one to the other. In practice, though, behaviour is seldom as simple as the description above suggests and a great variability in responses between individuals is often encountered.

There is a number of special cases of telotactic behaviour which have great biological interest. For example, to orientate the body so that its dorsal side is always up can be done by orientating at right angles to the light from the sky. This is the *dorsal light reaction* which is so important in determining the positional orientation of many kinds of animals; those animals which habitually have their ventral sides uppermost often show *ventral light responses*. An organism like the brine shrimp *Artemia* swims on its back, but if illuminated through the floor of the tank instead of from above, it swims dorsal side up directing its ventral side towards the light. Even in vertebrates where the inner ear supplies postural orientation information, this may be backed up by a dorsal light response. Fish with intact inner ears will swim tipped at 45° to the vertical if they are illuminated from the side; this is the compromise position between that which is correct according to the light sense and that correct for the gravitational sense. After destruction of its inner ears a fish will swim on its side when lit from the side, turning its back towards the light. In flying insects, the dorsal light reaction is an important means of controlling their flight path.

A number of animals move in nature at a definite angle to the sun's direction. Often this is a fixed angle with the result that, as the sun moves, so the direction in which they move alters correspondingly. Ants of various species have been shown to do this, most markedly in the early part of the summer. If they are trapped beneath a box when returning to their nest, they will take up a new direction on release. This is at an angle to their former direction approximately equal to the angle through which the sun has moved during their imprisonment. The inaccuracy introduced by the sun's movement during foraging time is small enough to be corrected by searching a small area for the nest on their return. It happens that this behaviour seems to be shown only by ants which are relatively inexperienced and are thus not foraging far afield. But as the ants mature, they use other methods of orientation. Specifically they begin to make allowance for the sun's movement so that they run in the same compass direction even after imprisonment. Such an allowance is typical of many animals from honeybees and sand-hoppers to birds and must involve a biological 'clock' to time them (see Chapter 5). In addition to using the sun as a compass, wood ant workers also learn visual land marks, as do many other animals (Chapter 8).

These forms of behaviour are not as rigid as they are sometimes assumed to be. For example, the sign of an animal's reaction to light may be changed or strengthened according to whether it is hungry or satiated, is desiccated or has had access to water, or whether it has been in the dark for some time or not. The reactions are generally typical only of a particular species, and genetically populations are polymorphic for this character, as they usually are for morphological ones. And, in addition, it is biologically necessary for animals to be released occasionally from behavioural responses which would otherwise trap them. Woodlice clustered under bark in darkness and humidity must emerge at some time to feed. It is not surprising to find, therefore, that as the temperature drops in the evening, their positive humidity responses weaken and they emerge to run about and find food.

The orientation of woodlice, and many other animals also, involves the phenomenon of *turn alternation*. If a woodlouse is put into a maze so that it has to turn through a certain angle in one direction, it will then tend to compensate at the next opportunity by turning through the same angle in the opposite direction. In its natural environment the animal will be able to maintain orientation in a certain direction in spite of obstacles in its path. This response is very easy to study in the laboratory, but is not very precise, so a number of turns must be measured and the mean taken before the effect becomes clear.

A nice example of the changing interplay of several orientation mechanisms is shown by the settling behaviour of winged aphids. These insects take off and fly upwards, being positively phototactic and attracted by the ultra-violet light from the sky. They fly for a time carried along by the wind. After a certain amount of flight activity their reaction to ultra-violet light changes and they become more attracted by those wavelengths of light which are reflected from plants. This leads them to land, where they test the leaves for suitability; if their probing proves them appropriate as food plants, they settle down inserting their stylets deeply into the leaf; if inappropriate, they take off again. Thus their dispersal over a reasonable distance is ensured by the postponement of settling behaviour.

The earlier workers concentrated on the responses of animals to very 'common' stimuli, such as light, humidity and gravity, but it is now that various animals can orientate with respect to ultrasonic sound, electric and magnetic fields with great precision. The African elephant fish (Mormyridae) and the South American knife fish (Gymnotidae) generate weak pulsed or oscillating electric fields from electric organs in their tails. Lissmann showed that non-conductors placed within their field cause the lines of force to diverge, and good conductors cause them to converge (as in a magnetic field). The fish are able to detect these distortions so that they can not only sense objects around them but can distinguish between similar sized objects on the basis of their conductivity. For at least some species it has also been shown that variations in the rate of discharge of the electric organ form a display used in communicating threat and the maintenance of dominance among a group of fish living together. The receptors are jelly-filled pits distributed through the skin of the fish. The knife fish and elephant fish have very poor eyesight and are found mainly in the large river systems of Africa and South America which are subject to great turbidity during the rainy seasons. Electrical navigation enables them to cope with this.

A somewhat analogous method of orientation involves the emission of high frequency sound and reception of the echoes; this has been well studied in bats and certain marine mammals such as dolphins.

Insectivorous bats emit pulses of sound in the ultrasonic range. The wavelength is sufficiently small for the sound to be reflected from quite small objects including insect prey. Neurophysiological studies have shown that the auditory centres in the bat brain are precisely tuned to detect small frequency shifts that occur in the echo (due to the doppler effect) and that are related to the distance between the bat and its prey. Some night-flying moths have hearing organs which are sensitive to ultrasound and the moths are able to orientate away from an approaching bat. When the bat gets very close, such evasive action is likely to be ineffective, and the moth makes a sudden rapid dive or similar manoeuvre which takes it out of the flight path of the bat. This evasive behaviour is mediated by quite simple ears, represented by a pair of tympana on the thorax or abdomen, each of which contains only two sensory endings that register vibration of the tympanum.

2.2 Navigation and homing

Navigation is a means of finding one's way to a certain point in space and therefore involves mechanisms of orientation. The migratory pathways of some birds, which move their centre of population from one general area to another with the seasons, may depend largely on light compass orientation. This is so in starlings, which can be shown in captive conditions to orientate with respect to the sun or an artificial light in a time-compensated manner. However, experienced starlings can reach their normal destination when displaced hundreds of miles from their flight-path, which suggests that they have some kind of conceptual map on which they know their approximate positions. It has been shown that some birds that migrate at night use the pattern of stars in the night sky as a compass, and possibly also as a map.

The homing ability of animals has baffled scientists for centuries. The Chinook salmon of the Pacific lives in the sea for up to five years and then returns to the stream it hatched in and spawns there. In doing this, it makes use of sun compass orientation and a highly developed sense of smell which enables it to discriminate between its home waters and those of other river systems. No explanation has yet been found for the navigation of the green turtle, which has feeding grounds off Brazil but returns to Ascension Island about 1000 miles away in the mid-Atlantic to breed. The behaviour of the Laysan albatross which breeds on Midway Island in the mid-Pacific is even more astonishing. Birds have been displaced to the Philippines (4000 miles) and to the west coast of the U.S.A. (3200 miles) and found their way back to Midway Island flying over a largely featureless ocean.

Many experimental studies have been made of the homing abilities of pigeons. Although they clearly make use of the sun, and of landmarks in their orientation they must use other cues as well, because they can still orientate homewards if the sky is overcast or if their internal clock has been experimentally reset by altering the phase of the light-dark cycle. They can even find their way home wearing frosted contact lenses. It is now believed that the earth's magnetic field is used by homing pigeons: this was previously discounted because pigeons with bar magnets strapped to their wings were able to find their way home, but recent experiments show that such pigeons orientate randomly under overcast skies. An artificial magnetic field can be induced in a pigeon's head by passing the current from a small battery through a copper coil (Helmholtz coil) round its head. In overcast conditions, pigeons were found to fly towards home when the magnetic field (i.e. the north-seeking pole) was inclined upwards, but directly away from home when it pointed downwards. However, this magnetic sense provides only an additional means of orientation and does not tell us how the bird is able to determine in which direction to fly. It is not yet clear whether the pigeon uses magnetic cues to supplement sun-compass orientation, or uses them only when orientation by the sun is difficult or impossible.

3 Sensory Factors in Behaviour

Many examples are known in which behaviour appears to be very automatic and inflexible. Examples in insect behaviour can be found in the writings of the famous French naturalist Fabre: for example, he joined two ends of a line of caterpillars of the pine processionary moth together so they walked continuously in a circle around the edge of a vase, each following a silk thread laid by those in front. They kept this up for seven days. Examples are also found in vertebrates of seemingly irrelevant behaviour. Male robins that have established a territory in the breeding season will attack and drive off other males, but will also attack a simple tuft of red feathers. Female turkeys will brood a stuffed polecat that contains a loudspeaker broadcasting calls of young turkey chicks.

3.1 Sensory filtering

In such stereotyped behaviour, animals can often be shown to be responding to only a very limited part of the information available to them through their senses. Neurophysiological studies have shown that the sense organs of some animals may respond only to a limited range of stimuli: for example, the ears (tympanal organs) of insects generally respond only to sudden changes in sound intensity: they cannot detect pure tones or discriminate between sounds of different frequency. Furthermore, sensory information may be processed on pathways to the central nervous system, and perhaps in sensory regions of the CNS itself, so that information that is not 'important' to the animal's response is filtered out. Where the sense organs are responsible for a great deal of sensory filtering the CNS or brain can be relatively unspecialized, as it is in many (but not all) invertebrates. In contrast, less filtering is done by our own sense organs; an immense amount of information about our environment is fed into the brain. This is particularly so for the visual and acoustic senses; we do not have such well-developed chemical senses, which is reflected in the scarcity of words in the English language to describe different kinds of odours. Our brains have the task of analysing all the information pouring in from our senses and determining, in conjunction with many other internal and external factors, appropriate responses. We can infer that the demands on the CNS must also be very great in the higher vertebrates. In the vertebrate brain it has been possible to map out the areas of the cerebral cortex according to their function. In animals such as dogs, rabbits and birds, virtually all the cortex is concerned with analysis of sensory input. In monkeys, such as marmosets, some of the cortex is freed from sensory function and given over to associative functions (correlation of sensory information, memory storage etc.) This trend is much further developed in primates such as chimpanzees, and culminates in the human brain, in which many of the purely sensory functions are confined to a narrow strip of cortical tissue, although sensory analysis does occur in other areas as well.

3.2 Responsiveness

We must remember that an animal's motivation, i.e. its readiness to respond, or its tendency to behave in particular ways, does not remain constant, but can be influenced by its hormonal state, how hungry it is, the time of day, preceding behaviour, and many other environmental factors. For a particular kind of behaviour, it is sometimes convenient to summarize the effect of all such variables in terms of a *drive*. It is important to remember that a drive is not a mechanism, but a tendency. The thirst drive depends upon factors including blood salinity, the time since the animal last drank, intake of dry food, etc. It is difficult to measure all such factors separately, and we rarely wish to do so, but we more often want to know the overall tendency of an animal to drink in a given situation, which can be measured behaviourally.

In a few cases, reduction in responsiveness has been shown to depend upon sensory stimuli. The presence of fresh eggs depresses sexual activity in male sticklebacks. Male field crickets will sing their calling and courtship songs if they have a ripe spermatophore. After mating, or after severance of the nerves to the spermatophore region, the insect ceases to sing these songs. The blowfly, *Phormia regina*, feeds intermittently until satiated. At that point, the crop is full and distended, and stretch receptors in the foregut signal the distension to the brain by way of a recurrent nerve: feeding then ceases. If the recurrent nerve is cut, this feed-back mechanism is destroyed and the fly will drink continuously until it may burst. The feeding 'drive' is influenced by input from chemoreceptors on the tarsi and on the tip of the proboscis. It is reduced as these adapt and stopped altogether by the activity of the feed-back from the crop region.

A number of authorities have suggested that the behaviour of animals is organized so as to ensure a physiological steady-state or optimum. The hypothalamus of the vertebrate brain is known to control responsiveness of this kind: it has centres which monitor levels of substances circulating in the blood. For example, one region controls feeding and is sensitive to glucose levels. If it is destroyed in rats they over-eat and become excessively fat. Another region monitors salt concentration, and injection of minute quantities of salt here stimulates drinking, and injection of water will stop a thirsty rat from drinking. There is also a temperature-sensitive region; electrical stimulation of part of this with micro-electrodes raises the body temperature and stimulation of another part lowers it.

Relatively long-term changes in responsiveness, such as those involved in various aspects of reproductive behaviour, may involve an interplay of hormonal and sensory mechanisms that can be very complex, and here the use of the drive concept or of simple models of motivation become useful.

3.3　Sign stimuli and releasers

Social signals, i.e. those stimuli provided by one member of a species which evoke certain behaviour in another member of the same species, were termed *releasers* by Lorenz. This term implies that the behavioural acts involved and the structures, scents or sounds implicated with them have undergone evolutionary modification into signals. Nowadays, the term releaser is commonly used in a more general sense to apply to any kind of signal to which an animal responds behaviourally. Tinbergen emphasized that in any releaser (a posture or display, for example) there were certain simple stimuli which are particularly relevant to the occurrence of a response, and he called these *sign stimuli*. 'Sign stimulus' and 'releaser' are now used interchangeably by many people, although this is often erroneous.

Sign stimuli can be visual, auditory or in any other sensory mode to which the receiver is sensitive. The visual ones are those which have been studied most, as they can be observed simply. Also they can be tested and analysed by using simply constructed models to offer certain stimuli only at a time. Thus, the aggressive behaviour of a male stickleback can be evoked by putting a mirror beside it so that it sees what is apparently an intruding male fish in it. The territory owner's reaction is to show strong aggression by standing almost on its nose in front of the mirror with his ventral spine nearest the rival raised (Fig. 3–1). A male siamese fighting fish will similarly display to his reflection, aligning himself parallel with the mirror, even if this involves swimming out of the vertical, and spreading his tail and dorsal fin. The male stickleback's behaviour can also be called forth by a dead fish mounted on a wire, provided that it is moved up and down. But a model, lacking fish shape, will be almost as effective, provided that its underside is coloured red. Thus the sign stimulus for this behaviour is a shape with red on its underside. The position of the red colour is important; it is not as effective on the back of the model.

That the posture of the threatening territory owner is decisive can be shown by catching him in a specimen tube. If he is in a tube which is narrow and, when it is on its side, prevents him taking up a vertical position, he will be less effective in keeping neighbours out of his territory than if he is in a wide tube in which he can stand on his nose (Fig. 3–1).

There are several other classical studies of visual sign stimuli which involved the general technique of presenting models to young birds in captive or field conditions. Lorenz showed that certain black shapes

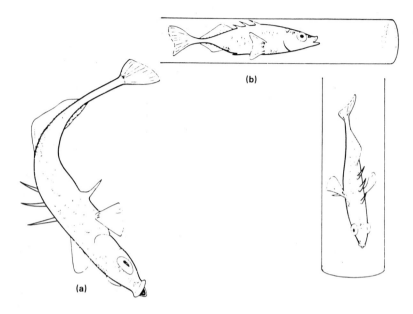

Fig. 3–1 (a) A stickleback in an aggressive posture at the boundary of his territory. (b) When in a horizontal tube the fish is unable to take up his threat posture though his spines are raised; in a vertical tube (below) his response is normal and now deters other males. (After TINBERGEN, N. (1951). *The Study of Instinct*. Oxford University Press, Oxford.)

flown overhead evoked escape responses in young turkeys. His findings, which will be discussed further below, were that a model with a short neck, wings and a long tail (all characters of a hawk silhouette) evoked the strongest responses but only if it was towed in the right direction. If the direction was reversed, so that the model appears to have a long neck and a short tail (like a goose, for example), then the model lost its effectiveness. Tinbergen and his associates found that gaping responses of nestling thrushes and blackbirds were directed, once their eyes were open, to a wide range of models presented about eye level. A black disc with a knob on was very effective. Rather similar experiments were done with herring gull chicks, which in order to obtain regurgitated food, peck at the red spot on the lower bill of the parent. Tinbergen and Perdeck used a cardboard model of a gull's head which they presented to chicks to test the importance of the red spot as a sign stimulus for the begging response. They found that a red spot on a yellow bill attracted most pecks, but spots of other colours were also effective providing that they contrasted with the colour of the bill. Surprisingly, a pointed stick with alternate red and white bands painted on it elicited even more pecks than a 'normal' model with a yellow beak and a red spot. The stick was acting as a *supernormal* stimulus, combining the essential stimulus elements of redness, colour contrast and elongation to an exaggerated degree. Oversize eggs may also provide a supernormal stimulus for brooding birds.

Whether there is any human behaviour that can truly be called innate has been hotly debated for many years, but some ethologists have claimed that certain aspects of non-verbal communication are common to all human cultures – including very isolated tribes – and are therefore innate sign stimuli. Eibl-Eibesfeldt filmed children who were born deaf and blind and found that their expressive movements of pleasure and anger were identical with those of normal children. The children had not been able to imitate others in a normal way and it therefore appears that environmental factors play a minor role in the development of their expressions. 'Eyebrow flashing', in which the eyebrows are rapidly raised and lowered, is ubiquitous in man during friendly greetings, although few people are ever conscious of its occurrence. Careful observation shows that there are many similar stereotyped movements and postures which can reveal a person's moods and feelings. These signals often have counterparts in the behaviour of animals (especially of primates) and have given rise to a number of very interesting, but often highly speculative theories on the origins of human behaviour (see for example, Morris' *The Naked Ape* or Eibl-Eibesfeldt's *Love and Hate*). Perhaps human facial expressions come nearest to the immediacy of animal signals, though they are affected by social convention and so forth. However that may be, the fact that animals signals are usually an indication of their physiological state gives a clue to their possible origin. The feather fluffing which is part of many a bird's aggressive or courtship postures is very like the way in which birds control their temperature, by raising their feathers a great deal to lose heat, not so far to conserve heat, or by flattening them to keep cool. If the physiological changes produced by the sight of a rival are similar to those which cause the feather movements of temperature control, then the state of the feathers is an indication of what is going on inside the animal, just as we know a person to be flustered when he blushes. Natural selection might then favour those whose feather movements were visually accentuated by the colouring of the feathers concerned. And also it might be advantageous for the feather movements to occur by the same amount whenever they are called forth. In this way the movements are turned into a signal by being ritualized.

This concept of ritualization is an important one because it suggests a way in which movements which are part of the motor repertoire of an animal can be changed into special signals. Cichlid fish furl their dorsal fins as they begin to move. This fin movement has become specialized into a signal to the young in a number of species. It is accentuated by the decoration of the fin with coloured spots in the African jewel fish (*Hemichromis bimaculatus*), so that the fin's movements are more conspicuous. In this species the signal causes the young fish to cluster around the parent. The intensity with which the fin is moved has also been heightened and standardized so that it is distinct from the fin's movements in swimming.

3.4   Change and development of responses

We have seen that the response to a stimulus depends upon an animal's motivation, but it may also depend upon the frequency with which it encounters that particular stimulus. Although a fighting fish will display to its own reflection in a mirror, if the mirror is left there for a week or two the number of responses it makes declines from day to day. This is a relatively persistent change in behaviour, a learning process known as *habituation*. When experiments were repeated with the 'hawk-goose' figure used by Lorenz, it was found that truly naïve turkey chicks did not discriminate between the figure in its two aspects. The birds used by Lorenz were kept in an area where geese and ducks were commonly seen flying, and had become habituated to the sight. This shows us that it is very important to take into account the previous experience of an animal before making assumptions about the innateness of its perception and behaviour.

In general, animals quickly learn to associate responses they make which are 'rewarded' in some way, by food for example, and tend to repeat those responses in an identical form. Indeed, this is the basis for training dolphins, circus animals and the like, to perform quite complicated tricks, and is a straightforward conditioning process. By careful experimentation, Hailman showed that the begging responses of gull chicks are very largely built up by such conditioning processes rather than being entirely innate as was formerly supposed. He first showed that chicks of the laughing gull would aim pecks at a model of the parent, which has a black head and a red beak. But newly hatched chicks showed very little discrimination and would peck at a wide range of models (Fig. 3–2), including those of the heads of other gulls, and rods held vertically and moved horizontally. Older chicks taken from the nest showed greater discrimination. Young herring gull chicks also responded to models of laughing gull parents, but after they had been in the nest for a few days they responded preferentially to a model of their own parent. Hailman then took newly-hatched herring gull chicks and gave one group food when they pecked at a laughing gull model, another when they pecked at a herring gull model and another group was fed before it pecked at a model. After two days of training, chicks of the first two groups responded more to the model with which they had been trained, even when not given any food.

Experiments such as these cast doubt on the interpretation of any experiments on the perceptual abilities of animals of which the developmental history is not known. Animals can sometimes change their perception rapidly; for example some birds are known to form, as a result of experience, a 'search image' of specific kinds of prey which they are then able to detect very quickly. Birds that find cryptically coloured caterpillars, for example, subsequently pay particular attention to the cues they provide and begin to search for them exclusively. They appear

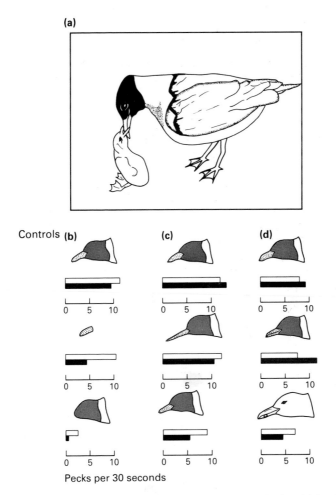

Fig. 3–2 (a) Pecking response of the laughing gull chick which elicits regurgitation of food from the parent. (b–d) Number of pecks given by day-old (white bar) and week-old chicks (black bar) to models of adult gull heads. (After HAILMAN, J. P. (1967). *Behaviour Suppl.*, **15**, 159 pp.)

to have acquired some kind of expectation of its appearance. This is very common in human behaviour; many years ago Aldous Huxley pointed out what every greengrocer knows, that oranges must be round and orange before people want them: when a person looks at oranges 'it is as though he looked at them through a stained-glass window representing oranges. If the real oranges correspond with the *beau idéal* of the oranges painted on the window, he feels that everything is all right'. Recent neurophysiological experimentation has shown that the eye and the brain are full of such 'stained-glass windows'.

D. H. Hubel and T. N. Wiesel recorded the activity of nerve cells in the visual cortex of the cat brain while stimulating the retina with spots or slits of light. They found projections from the retina such that a given cortical cell would fire action potentials when light filled a certain spot on the retina, or a certain slit-shaped area in the case of another cell. The orientation of the slit was important, some units responding to vertical bars of light, others oblique ones, and so forth. The retina is therefore not to be compared with a photographic film, but is organized into numerous receptive areas. Further evidence for this came from studies on the frog retina while exposing the frog to a variety of stimuli moved on screen providing a stationary background. Some cells showed no response to stationary black dots, but fired when one moved relative to the others: these were called 'bug-detectors'. Other ganglion cell units have been found which are thought to be adapted to the detection of other important visual features – slowly advancing predators, dimming of light, blue light reflected from water surfaces and so forth.

To what extent can sensory input modify the visual system so that it builds up its own expectation of the animal's world? C. Blakemore and S. Cooper addressed themselves to this question, and reared kittens in a restricted visual environment. In one group, individuals were housed in a cylinder in which they could see only vertical black and white stripes, and in the other they could see only horizontal stripes. Collars around their necks prevented them from seeing the outlines of their own bodies. After a period of five months the kittens were released from the apparatus and it soon became clear that their visual perception was abnormal. Those reared with horizontal stripes would play with a stick held horizontally, but ignored it when it was held vertically. Apparently they could not see vertical bars, and bumped into table legs. Kittens reared with vertical bars would not respond to horizontal rods. When the receptive fields of cells in the visual cortex were plotted it was found that most of the units had fields responsive to slits within 45° of the orientation, vertical or horizontal, that the animal had experienced. Similar experiments have since shown that kittens brought up in a spotted world have no line detecting brain cells at all.

Studies of the development of perception are crucial to the understanding of behaviour, and certainly undermine the usefulness of the concept of instinct. It may be that many of the particularities of human artistic cultures depend upon the stained-glass window effect: for example it has been found that North American Indians, accustomed to an environment rich in oblique lines provided by tepees, have difficulty in making discriminations which involve separating vertical and horizontal components of figures.

4 Courtship

4.1 Functions of courtship

The essential information passed in courtship is the sex and species of the performer, and his or her position, because the partners must make contact for mating and first they must find each other. The signals passed in courtship are usually so specific that they are barriers to interspecific mating and thus may be part of the process of speciation. The language of the signals is understandable only by a member of the transmitter's own species. Hence these signals become very elaborate and show a great variety differing in subtle details. The courtship patterns of the various *Drosophila* species, for instance, have common elements, but the order of these elements and the speed and frequency at which they are performed differ so that each pattern becomes quite unique to a species.

Many animals spend a great deal of time avoiding actual contact with other members of their species whatever their sex may be (see Chapter 8), but when mating is to take place this situation is reversed; contact with another member of the species is now an essential. Courtship itself very often involves overcoming the reluctance of one or both of the partners to reverse their tendency to withdraw from the presence of the other. In this process, not only is a suitable mate selected, but in many animals a pair bond is formed which may last for the season or for many years, depending on the species. This helps to ensure co-operation in looking after the young.

In general, readiness to respond to courtship signals and, indeed, the readiness to make them, depend upon internal factors, especially hormone levels (Chapter 6). A female receiving the male's signals does not therefore automatically respond to them. Courtship behaviour serves to synchronize the sexual maturation of a pair of animals, and ultimately ensures that sperm and ova are released together. Some animals, including members of the grouse family, form aggregations during courtship known as 'leks'. Males compete for favoured positions within the lek and gain the attentions of females only when they have attained those positions. In this, as in other forms of territorial behaviour, courtship signals both attract females and repel male competitors. Territory and courtship are intimately interrelated, but territorial behaviour is generally more marked in vertebrates than invertebrates. We will now take examples of selected groups of organisms to illustrate the nature and function of signals used in courtship and associated territorial behaviour.

4.2   Arthropods

A female *Drosophila* enclosed in a small cell, say, 1 in. in diameter and $\frac{1}{4}$ in. deep, with a male will go through the whole series of actions which is their courtship. This is best seen after the third day from emergence, for before this the female is relatively unreceptive; her receptivity rises to a peak at about $4\frac{1}{2}$ days, when it declines slowly to the tenth day. Details of the actions which form the behaviour pattern can be seen under a low power stereo-microscope. The main elements of the courtship will be found in most species, and certainly in *D. melanogaster*. A male approaches the female and taps her body with his fore tarsi, circling to one side of her so that he is now facing her side. A *D. melanogaster* male then opens and closes his wings moving them in a horizontal direction. During this behaviour, pulsed buzzing sounds are produced, although a very sensitive microphone is needed to detect them. The sound is detected by the antennae of the female, and the intervals between the pulses is a species-specific character. While orientating towards the female and vibrating its wing, the male periodically licks the tip of the female's abdomen with his proboscis. This is usually followed by attempted copulation in which the male mounts the female from behind. If she is unreceptive, she will kick him off or make other evasive movements.

In *Drosophila*, visual signals are relatively unimportant, and courtship will take place in the dark in many species. In many butterflies, the initial approach flight of the male is guided by visual stimuli. The male of the queen butterfly, *Danaus gilippus*, pursues the female in flight, overtakes her and dusts her with scent from two brush-like organs (hair pencils) which are everted from the end of the abdomen. A receptive female alights, and a ritual continues that culminates in mating (Fig. 4–1). The scent acts as an aphrodisiac. In many night-flying moths the complete courtship sequence, including initial attraction of the male, can be elicited by an attractive scent from the female (i.e. a sex pheromone – see Chapter 6).

Visual signals are strikingly developed in the fiddler crabs that are common on mudflats in tropical and subtropical areas. The males are very territorial and live in burrows. One of the claws is greatly enlarged in the male (Fig. 8–2) and usually very brightly coloured. The crab waves this up and down, standing near the entrance to the burrow, and this signal appears to have the dual function of attracting females and repelling other males. The trajectory and speed of movement of the claw vary greatly among species living in the same area (Fig. 4–2) and are species-specific characters. At night, and in certain other circumstances, the males make drumming noises by hitting their claws or legs on the substratum: these sounds seem to fulfil much the same function as the claw-waving.

Acoustic signals are of great importance in the courtship of many insects, apart from fruit-flies mentioned above. Male mosquitos are attracted to the female by the flight tone she produces. They detect this

Courtship of the queen butterfly

Female behaviour   Male behaviour

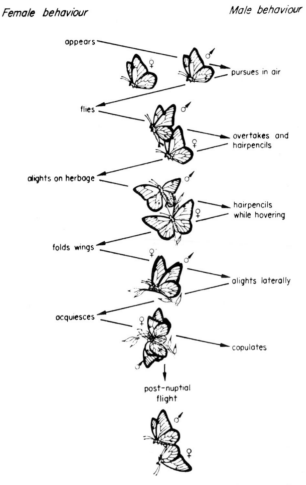

appears

pursues in air

flies

overtakes and
hairpencils

alights on herbage

hairpencils
while hovering

folds wings

alights laterally

acquiesces

copulates

post-nuptial
flight

Fig. 4–1   The stages in the courtship of the queen butterfly arranged as a chain of stimuli and the responses to them. (From BROWER, L. P., BROWER, J. V. Z. and CRANSTON, F. P. (1965). *Zoologica*, **50**, 18.)

with the aid of their fine antennae which are covered with whorls of hairs. Sexually motivated males will attempt to copulate with a tuning fork of the right frequency, and there is a record of an electrical transformer being put out of action because it was producing a hum that attracted thousands of mosquitos. Crickets, grasshoppers and cicadas have calling songs which the males commonly sing incessantly for long periods. European mole crickets, which live in burrows, use the burrow, which has

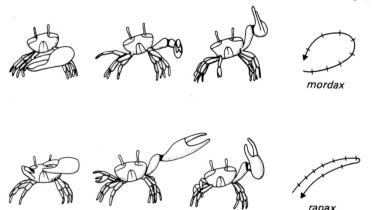

mordax

rapax

Fig. 4–2 The waving display of two species of fiddler crab, *Uca mordax* and *U. rapax*. The trajectory of the claw is indicated, in which the cross marks indicate jerks. (After SALMON, M. (1967). *Anim. Behav.*, **15**, 449–59.)

some similarities in form with the horn of a trumpet, as an instrument for broadcasting the song. Field crickets, which stridulate by movements of their wing covers, have several kinds of song. The calling song attracts females and is replaced by a softer courtship song when contact is made with a receptive female. A strident rivalry song is produced if another male approaches instead of a female.

4·3 The three-spined stickleback

If two male sticklebacks are put into the same tank in the spring each fish begins to dig a nest in the sandy bottom, picking up sand in its mouth and swimming a short distance away to spit it out. In the holes they have dug, each puts filamentous algae which it collects. The strands are glued together with kidney secretion, the fish wiping the underside of its abdomen and its cloaca over the nest. Then the fish makes a hole through the mass, forming a tubular nest. Meanwhile, and after this, each fish is defining its territory by attacking the other if it intrudes across its invisible boundary.

A territory owner will attack a trespassing fish by dashing at it with mouth open and dorsal spines erect. The appearance of this behaviour is, however, very dependent upon where the fish finds itself. On its own territory it will attack, but outside it rarely, if ever, does so. This can be shown by catching the owners of two neighbouring territories and putting each one into a test-tube of water. When the two tubes, held side by side, are put in the territory of fish A, A in its tube takes up the attack posture while B does not erect its dorsal spines and so forth. While in the territory of B, the reverse happens. The attacks rarely come to actual contact but if the trespasser does not turn tail, the owner will take up his threat posture. He positions himself vertically in the water, nose down (Fig. 3–1). He turns so that his flank is towards the intruder and he erects the ventral spine on that side towards the other fish. A fish will attack, and then threaten in this way, his own reflection in a small mirror stuck upright in the sand at the bottom of the tank.

In the spring, sticklebacks come into their courtship condition, the mature male fish having a red underside while the female is distinguishable more by her enlarged abdomen, swollen with eggs, than by her drabber colour. Even in an aquarium their courtship can be observed. A male remains in the vicinity of his nest (see later) until a female swims near. On seeing her, he performs what has been called a zigzag dance, for he swims in a series of short curved rushes around her (Fig. 4–3). If she is responsive, she curves her head and tail upwards. This signal to the male causes him to swim near and then down to his nest entrance ('leading'); the female follows him ('following'). At the entrance the male nudges the opening with his nose ('showing') and the female pushes past him to swim into the nest, remaining with head out of one side and tail the other. The male then pushes the side of her tail persistently, a signal which causes the female to release her eggs into the nest. This done she frees herself and swims away. Immediately the male swims into the nest and fertilizes the eggs. After this he commences parental care of the eggs, for the female takes no more part in the production of offspring. Here there are a sequential series of signals each of which can be shown by the use of models and so forth to evoke the next piece of behaviour from the partner culminating in the fertilization of eggs.

4·4   Birds

The courtship of birds uses visual signals, but sound signals are an important element as well. Much of the vocabulary of a bird's specific song is aimed at attracting a female and repelling rival males. Among many birds, territory is set up by individual males who leave the winter flock, and begin to defend an area by threatening other males often by means of an advertisement song from a song post situated somewhere in the territory.

Territory varies greatly in size and usage. Many sea birds which nest in dense aggregations defend only a small area around their nests: all the feeding is done elsewhere. Other birds maintain a territory to which they confine their feeding as well as their breeding activities. Yet others have territories that are used only for courtship and mating. Leks, which occur not only in birds, but also in insects such as damsel-flies and in some ungulates, are such examples. Finally, some birds and other animals that roost communally will defend their roost as a territory. Many animals do not confine themselves to their territories, but regularly traverse a (non-defended) *home range* that has larger confines. Within the territory itself,

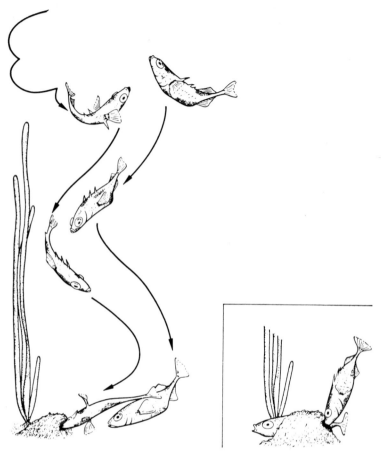

Fig. 4–3 The stages in the courtship of a pair of three-spined sticklebacks. (After TINBERGEN, N. (1951). *The Study of Instinct.* Oxford University Press, Oxford.)

they may spend most of their time in a smaller *core area*. Territorial boundaries rarely stay static for long, and there is much evidence that territorial behaviour plays a role in regulating population densities. Territories are often related in size to the food resources available, so in some birds they expand during the breeding season, when there are extra mouths to feed, and contract during the winter. If the population density increases, territories contract: they behave like elastic discs when they are pressed together. But there is a point beyond which they can be compressed no further, i.e. the territory holders will not give way, and then individuals may eventually be excluded and form part of a floating, non-breeding, population (which also happens in lek-forming species).

After some practice in observation, it is relatively easy to distinguish the courtship and aggressive behaviour of most common birds. In animals in general, an individual that appears to be engaged in aggression or threat (*agonistic* behaviour) adopts a posture which is more or less the antithesis of that shown by a submissive individual (Fig. 8–2). In birds (Fig. 4–4), this

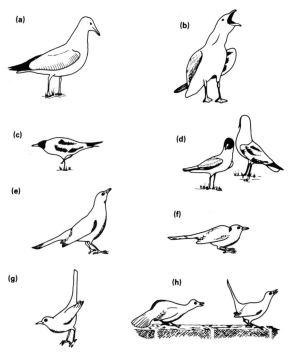

Fig. 4–4   (a) Upright threat posture in the herring gull, (b) oblique posture, (c) forward threat of the black-headed gull, (d) facing-away ceremony in a pair of black-headed gulls, (e) threat of male blackbird, (f–g) submissive postures, (h) courtship postures in a pair of blackbirds (male on the left). ((a–d) from TINBERGEN, N. (1959). *Behaviour*, **15**, 1–70; (e–h) from SNOW, D. W. (1958). *A Study of Blackbirds*. Allen & Unwin, London.)

often involves an upright stance with the head held high and the wings held a little way away from the body. In submissive postures, birds often reduce their apparent size by crouching, withdrawing the head and fluffing the neck feathers, so that their bodies appear more rounded in outline. An aggressive herring gull has an upright posture with the powerful bill directed slightly downwards towards its adversary. A forward threat posture is well-developed in the black-headed gull, where the bill is framed by the black mask of the head. A commonly-seen display of gulls on roof-tops is the oblique posture, given with a long drawn-out call with the bill wide open. This is a threat display, but serves also in the male to advertise the nest site to prospective mates.

The courtship behaviour of the male blackbird involves a posture clearly different from the threat posture, in which the head is stretched forward with the beak open, and the feathers are raised on the rump (Fig. 4–4). On a branch, he bows up and down in front of a female; even on the ground he also bows. If the female is paired already, she attacks the male. When she has formed a pair bond, she elicits copulation after courtship. She stands with her legs extended, beak stretched upwards and tail pointing skywards; her feathers are sleeked, and she utters a soft call which can only be heard close by.

A necessary prelude to courtship in gulls is appeasement behaviour, which reduces aggressive tendencies between the bird on the nest site and the mate which is at the same time an invader of the small territory. In black-headed gulls this takes the form of a 'facing away' ceremony, in which the two birds align themselves parallel to one another and then turn their heads away, so diverting the threatening beak from the view of the partner (Fig. 4–4). Appeasement displays of related kinds are common among gulls.

Some birds are renowned for their elaborate colouration and displays: birds of paradise are obvious examples. So are the bower birds, in which the males build a bower of twigs and other materials and then select brightly coloured objects – flowers, seeds, etc. – which they lay out in a carpet near to the entrance of the bower. This is a courtship display also.

It is sometimes possible to trace stages in the probable evolution of courtship displays by comparing the behaviour of closely-related species. This has been done for waterfowl, in which it can be surmised that the various movements involved in, for example, drinking, preening, and flight take-off, which are likely to occur more often when the birds are in a conflict situation such as courtship provides, have been subject to selection and become ritualized (p. 19) to form part of the courtship display. In this, they have lost their original function, and their neurophysiological mechanisms have evidently become tied to the new behaviour; a process known as *emancipation*. A simple example is provided by preening movements which seem to have undergone a process of progressive elaboration, starting with the garganey drake in which the beak is run through the wing feathers, proceeding to the mallard, in which the beak exposes a bright blue feather in the wing, and culminating in the mandarin drake, which simply points its bill towards a conspicuous bright orange sail feather in the wing. In ritualization, the intensity of movements tends to be increased and standardized. Ritualized displays tend to be repeated rhythmically, and the speed and timing differs from that in the behaviour patterns from which they are derived. They are usually associated with conspicuous body features. All these are aspects that make for clear and unambiguous signalling.

5 Rhythms

Rhythms can be traced in most aspects of behaviour. Courtship and nesting in birds is a familiar example of an annual rhythm, in which activity commonly reaches a peak in the spring in temperate latitudes or with the beginning of rains in the tropics. The brood then begins to mature when there is a plentiful supply of insects or seeds for food. Sometimes annual rhythms coincide with lunar periods, especially in marine animals in which activity is related to the tides. Palolo worms show a remarkable swarming behaviour confined to brief periods of the year. They develop body segments filled with gametes and swim to the sea surface in enormous numbers. The Samoan Palolo, which is collected as a delicacy by the Polynesians, swarms almost entirely on one day of the year at the beginning of the last lunar quarter, the average date being 27 November. The timing mechanism for this is not understood, although it is better known in another marine animal, the midge, *Clunio maritimus*. The immature stages live in the sea, but the adults can emerge, mate and lay eggs only at times of extreme low water, which occur twice in the lunar month. Laboratory experiments have shown that if cultures of the insect are exposed for four nights to artificial light of about the same intensity as the full moon, at intervals of thirty days, the adults emerge during the following four days and at bimonthly intervals thereafter.

We are mainly concerned in this chapter with rhythms having periods of about 24 hours – known as circadian rhythms – but there are some rhythms which are much shorter and cannot be related to environmental cycles of light and dark, temperature or whatever, but which depend upon pacemaker properties of nerve cells. Rhythmic walking or flying movements of invertebrates come into this category. The desert locust for example, flies at around seventeen wingbeats per second: the rhythm is generated within an interacting network of nerve cells in the thoracic ganglia. The rhythm is only partly affected by sensory inflow: cutting stretch receptors at the wing bases causes the wingbeat frequency to drop by about one-third. Many animals show short-term rhythms of feeding behaviour. Polyps of the soft coral, *Heteroxenia*, open and close independently 35–40 times per minute. The lug-worm, *Arenicola marina*, lives in a U-shaped burrow in mud or sand and has a feeding cycle during which it ingests sand and detritus, with a period of about seven minutes. At forty minute intervals it irrigates its burrow and may move backwards to defaecate, adding to the cast at the exit of the burrow.

A worm can be persuaded to burrow in mud between two glass plates clamped together with a U of rubber tubing between them acting as a spacer. The whole lot is immersed in sea-water so that it covers both ends of the burrow. A float is arranged over the tail end of the burrow and attached to a lever writing on a smoked drum. The worm's movements cause the water level to change and thus the lever is moved. So, as G. P. Wells said, 'The worm writes its autobiography'! This demonstrates the regularity of the movements which continue even though the worm has no tidal or light cycle impinging upon it. This rhythm seems to have a pacemaker in the nervous network investing the pharynx, for an isolated pharynx shows contractions of the same period as the whole worm's movements.

5.1   Temperature-dependent rhythms

Rhythms that are determined by external factors are known as *exogenous*, while those involving some kind of internal clock or pacemaker are called *endogenous*. It is usually very difficult to decide whether a rhythm falls into one category or the other and most seem to have both endogenous and exogenous components. Some aspects of behaviour are very much dependent upon temperature, particularly in 'cold blooded' animals, so that any endogenous factors appear to be of lesser importance in determining what an animal does. Many insects and reptiles begin their day with a period of basking during which they gain heat from the sun by exposing a maximum area of their body towards it. As the body temperature rises they become capable of more complex activities.

Lizards commonly show shuttling behaviour, alternating between sunlit and shaded areas. Early and late in the day they bask for relatively long periods, but towards the middle of the day basking periods get shorter and the rate of shuttling increases. Similar behaviour also occurs in insects such as cicadas. *Magicicada cassini* seeks shade at a temperature of about 32°C but reverts to sunlit areas when the temperature falls to 25°C or below. The insect is capable of feeding and walking over a much wider temperature range (10–40°C) but courtship and singing occur only between about 25° and 35°C.

Insects also exploit temperature differences in a stand of vegetation. The silver-washed fritillary flies to the top of trees when temperatures are low and basks in the sunlight with its wings widespread. As it gets warmer, it reduces the area of its wings exposed, and eventually flies lower down in the vegetation, retreating into the shade with its wings closed and abdomen drooping when its body temperature rises too high. Its courtship behaviour occurs at different times from day to day according to the ambient temperature and the ability of the insect to attain its optimum temperature. Moths and some other night-flying insects cannot easily gain heat from the environment, but generate it by contracting the flight muscles rapidly to vibrate their wings in behaviour known as 'shivering'.

Social insects employ various methods to maintain a steady temperature within their nests. Air circulation systems are developed in the mound nests of some termites. The temperature within the hive of the honeybee is kept within one degree of 35.5°C. When the temperature falls the bees cluster together, so conserving heat developed metabolically. When it rises above the optimum, worker bees bring water which they spread on the surface of the comb. They assist evaporation of the water by fanning with their wings, and bees at the hive entrance also fan, driving air through the hive.

5.2   Rhythms independent of temperature

Most of the circadian rhythms that have been studied are scarcely affected by temperature at all. Animals have characteristic activity periods which often differ among closely related species. A good example of this is seen in mosquitos. They begin to bite around sunset, and in East African forests it is possible to be bitten by a number of species in sequence in the space of a few hours, some of which have very sharp peaks of activity lasting for only about twenty minutes (Fig. 5–1).

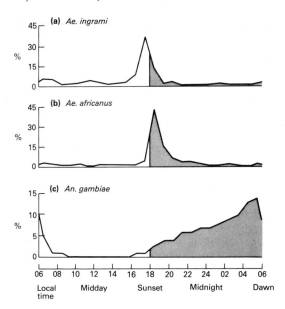

Fig. 5–1   Feeding cycles of three mosquito species (*Aëdes* and *Anopheles*) in East Africa. (After GILLETT, J. D. (1971). *Mosquitos*. Weidenfeld and Nicolson, London.)

In insects, also, circadian rhythms of adult emergence (eclosion) from the pupa are common, and have been studied in insects such as *Clunio* (already mentioned), *Drosophila*, and silk moths of the family Saturniidae. A culture of *Drosophila* kept in constant dark conditions has no eclosion rhythm: insects emerge throughout the 24 hour period. But if the culture is exposed to a 24 hour light-dark cycle, adult emergence occurs 24 hours later and continues at roughly 24 hour intervals thereafter. The light signal has synchronized the emergence rhythms of individuals in the population and is an example of a *Zeitgeber* (German: time indicator). A Zeitgeber is not always a light stimulus, it can sometimes be a temperature change, or, in marine animals, a change in hydrostatic pressure or turbulence. Quite commonly, animals synchronize their activity rhythms with a light-dark cycle having a 24 hour period. Mice, for example, are active mainly during the first part of the dark period, while chaffinches are mainly active during the first part of the light period. If there is an abrupt shift in the light-dark cycle the activity periods do not change in a single step in many animals, but gradually rephase over several cycles. This process is known as *entrainment*.

If an animal is kept in constant light or constant dark, circadian activity rhythms usually continue for a few days or weeks before the animal becomes arrhythmic. Such 'free-running' rhythms have periods which are a little more or a little less than 24 hours. In general, the period of day-active animals shortens in constant bright light, while that of night-active animals lengthens, and vice versa. Thus mice kept in constant bright light lengthen their activity period, while starlings shorten it. People kept in underground bunkers without time cues also show free-running rhythms, developing a 33 hour 'day' of sleeping and waking, and a 25 hour 'day' of peaks in body temperature and urine secretion. It appears that our bodies have a number of more-or-less separate internal clocks controlling circadian rhythms.

If rhythms persist in constant conditions, it is tempting to think that they must be endogenous. It is, however, almost impossible to ensure constant conditions: light and temperature are only two cyclical variables in the normal environment, and many other factors, such as barometric pressure fluctuations and fluctuations in the earth's magnetic field are known to affect animal behaviour and may provide Zeitgebers for activity cycles. However, most of the evidence points to the endogenous nature of most circadian rhythms. For example, genetic differences have been found in the length of the free-running period in different strains of mice, and in populations of *Clunio* from different parts of the European coastline. Translocation experiments, in which animals have been taken across time zones, also provide strong evidence for endogenous clocks. Renner trained honeybees in an experimental flight chamber to come to a food dish at a certain time of day in Paris. He then flew the bees to New York and tested them in an identical chamber in an enclosed room. They came to the food dish 24 hours after the Paris training time, although they had undergone a time change of several hours. They did not rephase their activity to local conditions, but when they were trained outdoors in New York and flown to California overnight they first foraged at New York time but gradually rephased their activity to Californian conditions when allowed to forage in the field there. Modern transatlantic travellers are

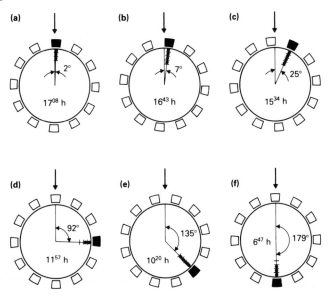

Fig. 5-2 Choices of feeding position made by a starling trained to the west with an artificial sun. Large arrows indicate the direction of the artificial light. The empty blocks indicate food hoppers; the black block indicates the 'expected' choice of food hopper assuming that the bird shows time-compensated sun-compass orientation. (After KRAMER, G. (1952). *Ibis*, 94, 265-88.)

familiar with 'jet-lag' resulting from their circadian rhythms falling out of phase with the daily cycle at their destination. Rephasing takes from a few days to several weeks, depending on the individual.

There has been an intensive search for a circadian clock mechanism in insects during the past decade. Despite the ease with which insects can be deprived of their organs and body segments and yet still remain active, many conflicting results have been obtained. It now appears that there may be a number of interdependent clocks controlling physiological, behavioural, growth, and cellular processes in insects, as there are in man, and the search for a unitary clock is doomed to failure.

5.3 Circadian rhythms in orientation

There is no doubt of the biological usefulness of circadian rhythmicity in sun-compass orientation; the sun can be used as a compass only if animals have some means of allowing for its movements during the day. This ability is well-developed in animals that migrate over long distances, such as some butterflies, fish, turtles, and birds (see Chapter 2). It is also present in honeybees, which use the sun as a compass in foraging, and in invertebrates such as *Velia*, the pond skater, and *Arctosa*, the wolf spider,

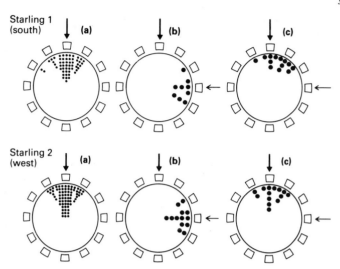

Fig. 5–3   Choices of feeding position in two starlings, one trained to the south, the other to the west (a). After 12–18 days in an artificial light cycle 6 h behind local time choices are 90° out of phase (b), but return to normal after 8–17 days in a normal day (c). (After HOFFMAN, K. (1971). *Proc. int. Symp. circadian rhythmicity, Wageningen*, 175–205.)

which forage on the water surface and use the sun for orientation in escape movements towards the nearest bank. Sand-hoppers, *Talitrus*, live on the upper seashore and orientate down the beach and feed near the water-line. They can orientate during the day by using the sun, and also at night, by using the moon as a compass.

Dependence of light-compass orientation on a circadian clock can be demonstrated by keeping the animal in an artificial light-dark cycle out of phase with normal conditions. Kramer published the results of some elegant experiments in 1950 in which he showed that starlings kept in an hexagonal aviary during the migratory season became restless and chose perch positions where they were orientated towards the normal direction of migration. When the birds could only see the sun reflected by a mirror, they altered their positions and so were clearly responding to the sun's rays and not to the earth's magnetism or some other factor. Later, Kramer showed that starlings could be trained to look for food in a certain compass direction, and an artificial light source could be used to substitute for the sun. With the light fixed in position, the starlings behaved as if it had moved like the sun and twelve hours after training orientated to the food dish diametrically opposite (Fig. 5–2). In similar experiments, Hoffman trained starlings to search in a particular direction and then kept them for 12–18 days in an artificial daylight cycle six hours behind local time. When the birds were tested in the sun they then orientated at around 90° to the training direction (Fig. 5–3).

6   Hormones and Pheromones

6.1   Hormones

The physiological nature of 'drive' is a matter of considerable dispute, but that hormones are implicated in one way or another in reproductive drive is indisputable. Unquestionably much of the special behaviour is made possible by the presence or absence of hormones at some particular point in the body. An animal's – or a person's – mood may be determined by the particular hormones which are circulating in the body, rendering the animal peculiarly sensitive to certain stimuli from the environment, so that, say, those evoking reproductive behaviour are most likely to produce appropriate activity which at another time of the year, when those hormones are not there, will be unlikely to appear even in the presence of the same stimuli.

The implication of hormones in the sexual cycle of vertebrates has been particularly well studied. In mammals, for example, the body of the female must be prepared physiologically so that the result of the mating, the fertilized ovum, may be dealt with appropriately. The increasing daylengths in spring are known to affect birds leading to the enlargement of their gonads. This result is produced by the light stimulus received by the eyes acting upon the brain and the pituitary in turn. The increased size of the gonads represents increased output of sex hormones, which cause reproductive behaviour such as territorial defence and the appropriate song, the marks of increasing aggression which make male birds isolate themselves from the communal flock in which they have passed the winter. In the song sparrow as the season goes on, and gonad activity increases, the air temperature at which song is inhibited decreases so that song is more easily evoked; a result which could equally be interpreted as an increase in drive, making it more difficult to suppress the singing. In the autumn there may be a slight increase in the gonad activity, smaller than that in spring, but sufficient to produce territorial song and the beginnings of male behaviour, without all the courtship behaviour which follows in the spring. This shows how closely linked gonad activity and behaviour are.

Though the breeding cycle of birds appears to be entirely determined by the environment (exogenous), namely by the influence of light on the brain, there is evidence that they must have an internal clock which enables them to detect changes in daylength. Some species, in fact, such as the short-tailed shearwater, show a seasonal gonadial change in a constant light-dark regime, and collared doves do the same in a variety of photo-periods, so these birds, at least, must possess an annual physiological clock of some kind.

The maternal behaviour which in many vertebrates follows mating is similarly controlled by hormones. Doves, for example, feed their young on 'milk' produced as a secretion from the walls of the crop. For the milk to be produced, prolactin from the pituitary must be circulating in the blood, while participation in the incubating of the eggs stimulates production of the milk. The importance of the activity of incubation was shown in experiments with birds injected with prolactin. The injections were sufficient to develop the crop and also to suppress sexual behaviour for seven days. Despite the fact that the crop was in the state to produce milk, faced with hungry seven-day-old squabs, birds with no previous experience of incubating and feeding failed to feed these young, though they were making begging movements and sounds. Even birds which were experienced reacted only after sixty minutes, much longer than the normal time that an adult takes to respond to such begging behaviour from squabs which it has incubated. The age of the young seems to be important, for newly-hatched squabs seem more potent in evoking parental behaviour; even inexperienced doves injected with prolactin will feed hatchlings.

Equally, male sex hormones can bring about aggressive behaviour in young animals. Young cockerels, at an age when they would not yet normally have developed aggressive tendencies to others, will show all the aggressive behaviour of a much older bird when injected with male hormones. Position in a peck-order hierarchy in birds is strongly determined by the level of sex hormones in the blood; the higher the level the higher the position.

Though there is so much evidence for the connection between mood and hormones in vertebrates, it has proved much more difficult to show a similar connection in invertebrates. A number of hormones are known in crustacea and insects, to take the most obvious examples, which, however, are implicated in growth and development processes. They are only marginally involved in determining behaviour. For example, the light responses of the eyed hawk moth caterpillars alter when the level of the juvenile hormone is changed; the juvenile hormone causes increased photopositive behaviour, while the moulting hormone does the opposite. Some of the special behaviour connected with moulting and metamorphosis in these animals may well be determined by the hormones which are also bringing about the developmental changes.

In some insects, but not all, development of sexual behaviour in adults, and its maintenance, depends upon hormones secreted by the *corpora allata* (which also secrete juvenile hormone). In wild silkmoths of various species, the *corpora cardiaca* have a major role in the determination of behaviour. An 'eclosion hormone', secreted from the corpora cardiaca and brain towards the end of development, initiates the sequence of behaviour necessary for emergence from the pupal case and escape from the cocoon, and also 'switches on' nervous mechanisms that can generate the full repertoire of adult behaviour. One of these is 'calling behaviour' in the female: exposure of scent glands from the tip of the abdomen, used in attracting males (see below). Calling occurs only during a limited period of the light-dark cycle, and is elicited by a calling hormone released from the corpora cardiaca. After mating, sperm in the bursa copulatrix of the female causes a release of a factor into the blood which operates a switch to 'mated behaviour' in which she ceases to call. An oviposition hormone is now released from the corpora cardiaca, again at certain phases of the light-dark cycle, and this increases the general activity of the female and induces her to lay large numbers of eggs. Hormones can thus have two kinds of effect in these moths; they can either release behaviour, such as calling, oviposition, or eclosion, or they can have a more general modifying effect, predisposing to adult, rather than pupal behaviour, and mated, rather than virgin female behaviour.

6.2   Pheromones

Odour plays a relatively small part in our lives, and it is difficult for us to imagine animals living in a landscape of odours that may be as important to their behaviour as vision is to ours. Many animals use scent for finding food, or a mate, for spreading the alarm, marking territory, and so forth. Chemicals that an animal secretes and which affect the behaviour or physiology of another of the same species are called *pheromones*. This term has the same Greek root as the word hormone: when it was coined in 1959, pheromones were mainly thought of as single compounds acting as external messengers but otherwise analogous to hormones that act within the body. However, in contrast with hormones, pheromones are usually relatively volatile substances of low molecular weight, they commonly occur in multicomponent mixtures in which even minor components may be of great importance in communication, and they may sometimes also have an action (defensive, for example) on other species. In the latter aspect they can no longer be termed pheromones.

The first pheromones to be isolated and identified were anal gland secretions of animals such as the civet cat (civetone) and the musk deer (muscone); these substances were available in copious quantites and were used as bases in the perfume industry and their biological significance was largely ignored. Later, the sex attractant pheromones of Lepidoptera were studied, a much more difficult task because usually only a few nanograms of these pheromones are present in each insect. It has been recorded for hundreds of years that the caged females of some moths attract males from long distances; a phenomenon that entomologists call 'assembling'. This depends on a scent produced by the female from eversible glands at the tip of the abdomen. The first such scent to be isolated and identified was a long-chain alcohol named *bombykol*. To obtain this, A. Butenandt extracted the abdominal tips of 250 000 female moths. Since that time (1959) chemical techniques have advanced considerably and pheromones can usually be detected and isolated (although not necessarily identified) from only a few insects by gas chromatography. It has also been found that animals of many kinds, including mammals, fish and amphibia, employ pheromones as a means of communication. There is even evidence of their use in man in the form of skin odours and volatile sexual secretions.

The social insects are very dependent upon chemicals for both communication and defence. A number of pheromones is involved in honeybee behaviour. The best known of these is the 'queen substance' produced by the mandibular glands of the queen, which suppresses ovariole development in the workers. This is disseminated throughout the colony by the constant licking and grooming (*trophallaxis*) of the queen by the workers and of the workers by sister workers. If the queen dies, or if the colony becomes large, the amount of queen substance in circulation diminishes. In consequence, the workers begin to build queen cells and rear new queens. 'Queen substance' in fact contains many chemicals; two of the most important are 9-oxodecenoic acid and 9-hydroxydecenoic acid, and outside the hive these are known to have different and separable effects. An additional queen in a large colony may leave on a mating flight. Then, the 9-oxodecenoic acid acts as an attractant for drones when the queen is flying above the ground in a wind, indeed, drones can be attracted to a phial of the acid flown from a balloon. The original queen usually leaves the hive with a swarm of workers towards the end of the season, and daughter queens normally do the same after they have mated. Swarming bees that have lost their queen are attracted upwind by the odour of 9-oxodecenoic acid, but will not settle and form a stable cluster unless the hydroxydecenoic acid is also present. Honeybees produce a pheromone from a gland associated with the sting which attracts other workers and makes them aggressive. They produce an attractant from a gland near the tip of the abdomen which helps to guide bees landing near a good food source or near water they can drink. Another secretion is used to mark the nest entrance, and it is to be expected that still more pheromones will be discovered in the coming years.

Many ant species, but not all, make use of chemical trails in foraging. Trail pheromones can be amazingly potent, although the time for which they persist when not renewed varies among species. Leaf-cutting ants make long trails to plants they exploit. One component of the trail pheromone of the North American species, *Atta texana*, has been found to be active at levels down to 10^8 molecules per centimetre of trail. Theoretically, 0.1 of a microgram would be sufficient to draw a detectable trail around the world. Leaf-cutting ants, like many other species of ant, have mandibular glands producing a pheromone that is used to communicate alarm. Alarm pheromones are usually very volatile and so spread alarm rapidly, but at the same time do not persist if the emergency is short-lived.

A substance is released from the skin lesions of wounded minnows which acts as an alarm pheromone, elicting frenzied escape responses in other minnows in the same water. Pheromones of other kinds are known in a variety of fish, including blennies and catfish. Individuals of the bullhead catfish can recognize one another on the basis of an odour connected with skin mucus. The fish can live communally, and clustering results in the production of a pheromone that inhibits aggression. This was shown by circulating water from a tank containing communally living fish to one containing a pair of territorial bullheads. After seven days the aggressive behaviour of the latter was greatly reduced.

The use of urine and faeces as markers for trails or territories is very common among mammals: the domestic dog being an obvious example. The hippopotamus uses faeces in this way, spreading them by a rapid fan-like activity of its short tail. Rabbits urinate on their family members giving them a specific odour which aids recognition. Faeces are coated with a pheromone from anal glands and then used for marking the territory of the rabbit. Within the territory, local features, including the entrance to the burrow, the females and young, are marked with a secretion from a gland under the chin. The size of the chin and anal glands are under hormonal control and are greater in more dominant individuals of a social group and greater in males than in females.

A sexually aroused male pig produces a rich supply of saliva containing hydroxysteroids from the submaxillary glands which are also responsible for the taint of boar meat. This pheromone is now used commercially in artificial insemination practices. It is used as a test for a sow in oestrus, which shows an appropriate behavioural response to the odour.

Chemical communication has been studied in some detail in the black-tailed deer, which has at least six separate pheromone systems (Fig. 6–1). A gland on the forehead is used for marking objects in the home range. A metatarsal gland produces a garlic-like odour which acts as an alarm pheromone. The tarsal gland produces a scent which is peculiar to a particular individual and also gives indications of the sex and approximate age of the deer. Urine has a variety of functions; the urine of females attracts males, adult deer urinate during agonistic encounters, and fawns urinate when they are distressed.

We have seen that the queen substance of the honeybee can have physiological effects upon the reproductive system of bees. A similar effect has been found in the desert locust, *Schistocerca gregaria*. The adult males are mainly pink in colour but become yellow as their gonads mature. If the immature males are kept separately from one another they mature faster in the presence of mature ones or of cotton wool pads soaked in an ether extract of mature males. This shows the existence of a maturation pheromone which is probably sensed by receptors on the antennae. The brain is then thought to trigger activity of the corpora allata to produce a maturation hormone. In the Red Sea area, sexual maturity of desert locusts can often be retarded for several months until just before the rains break. This is induced by fragrant desert shrubs bursting into flower just before the rains and releasing into the air

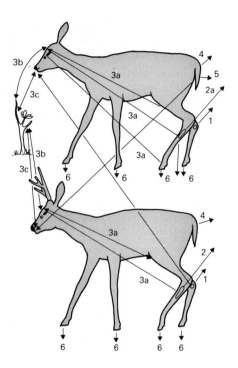

Fig. 6–1 Pheromone pathways in the black-tailed deer. The scents of the tarsal gland (1) are used in mutual recognition and that of the metatarsal gland (2) communicates alarm. Scents of the metatarsal gland are put on the ground (2a) when the animal reclines. Pheromones from the leg glands are transferred to the forehead by rubbing (3a) and are then possibly transferred, together with the forehead gland secretion, to twigs (3b). Marking is done mostly by males in their home range and females sniff and lick the marked twigs (3c). Female urine (5) is attractive to males, and the tail gland secretion (4) is sniffed by deer when they are excited. Interdigital glands leave scent on ground (6). (After MÜLLER-SCHWARZ, D. (1971). *Anim. Behav.*, **19**, 141–52.)

terpenaceous scents (including those of frankincense and myrrh) which have a pheromone-like action on the locusts.

Pheromones which have an action on the endocrine system are also known to occur in mammals. Oestrus is suppressed in groups of female mice, but can be triggered or accelerated by the presence of urine of a male mouse. The urine of an unfamiliar male can also prevent implantation of the ova in a newly-mated female, but the urine of a familiar male does not have this effect. Such phenomena may be fairly widespread in mammals and we must learn more about them if we are to understand mechanisms of social behaviour and population fully.

The study of pheromones will also help us in understanding our own behaviour. For example synchrony and suppression of the menstrual cycle has been found in women students living communally in a hall of residence, and may have a pheromonal basis. Women are also more sensitive than men to various odours including the smell of other peoples' clothing and of steroidal compounds given off from the skin or in sweat.

Pheromones and similar natural product chemicals may well become increasingly important to the future of mankind by offering new or improved methods for the selective control of pests. Effective control of some insect pests of agriculture has already been achieved in limited trials using sex pheromones as attractant baits for traps. Another promising method has involved the use of chemical inhibitors of sex attractants which prevent males from finding and mating with receptive females.

7 Learning

7.1 The nature of learning

It is traditional to discuss learning as if it were set apart from all other forms of behaviour. This is partly because much of experimental psychology is concerned with ways in which new behaviour patterns arise and in which problems are solved by animals and man. In the past, comparative psychologists have devoted considerable efforts to the study of learning abilities of laboratory rats but in more recent years have diversified their interests to include experiments on animals such as siamese fighting fish, cockerels, and chimpanzees. Hence, psychologists and animal behaviourists are learning from one another, and finding that animals show many kinds of adaptive modifications of their behaviour, both in the laboratory and in the field.

Learning is usually defined as an adaptive change in the behaviour of an individual as a result of experience. It involves the central nervous system (at least in metazoa) and is more or less permanent. This definition must be hedged with qualifications. In the first place, it is taken by some people to mean that behaviour can be divided into parts which have their mechanisms formed by the environment and part (instincts) which do not. Such a dichotomy is unreal (p. 6), and there are some adaptive changes in behaviour which most people would not consider to be learning. For example, if a negatively phototactic insect is kept for a long period in the dark it may become positively phototactic. As a result of experiencing an inverse light-dark cycle for a period of a week or more many animals will rephase their activity periods and keep them that way for at least some days under constant lighting conditions. In the very young animal, behaviour patterns may develop in parallel with the growth of muscles, nerve connections to them, and so forth. It may thus seem that a tadpole is 'learning to swim' or a bird 'learning to fly' because it appears to improve as a result of practice, when in fact the changes depend upon the growth of effectors and the nerve pathways to them. Such maturation of behaviour patterns depends, in ways that are still little understood, on feed back from the environment. Recently, there has been a growing interest in the development of sensory aspects of behaviour (Chapter 3) which has led to the realization that experience before hatching, or before birth, is important to future behaviour. The retina of the chick begins to respond to light at about eight days before hatching, and if the eggs are exposed to a flashing light the chicks show a greater tendency to approach one after hatching. It has also been shown that young ducklings of at least some species learn features of the maternal call before hatching.

Experience which results in adaptive changes in behaviour can therefore be very difficult to define. This apart, the various phenomena that are commonly described as types of learning may well have different underlying neurophysiological mechanisms. Within the nervous system, some sort of memory trace is evidently established on which the animal can call later. There is much evidence that learning begins with short-term memory. During the first hour or so after training, mammals can have their memory destroyed by shock of various kinds, but after this period shock treatment has no effect. An octopus can learn to take a crab offered with a visual stimulus, like a square, and avoid one accompanied by some other kind of symbol. Normally, such learning lasts for days, but if the connections of the vertical lobe with the rest of the brain are cut, learning lasts hardly more than two minutes. The operation seems to remove the long-term memory store. Short-term memory is thought to be formed by temporary changes in the synapses of nerve cells, mainly of an electrical nature. Long-term memory, on the other hand, is believed to involve biochemical changes in nerve cells, perhaps accompanied by the development of new contacts between cells.

One apparent test for the role of learning is the isolation test where the animal is reared solitarily away from members of its species and unable to see, hear or smell them. This avoids experience which the animal could gain from the other animals, but it cannot prevent the animal experiencing its own movements, or its own noises disorganized though these may be. To prevent this would require that the animal be anaesthetised, deafened and so forth, procedures which hardly leave the animal a fit subject from which sensible conclusions can be drawn. In fact, this self-experience seems to be very important in the formation of many behaviour patterns, so Kaspar Hauser experiments, as they are called, are not as conclusive in this context as many people once took them to be, but they remain valuable and they are indispensable in studies of social deprivation.

7.2   Habituation

All young animals will react to strong stimulation on the whole, by retreating from it. A loud noise, or a flash of light causes a young bird to crouch or a kitten to run away. But gradually if such a stimulation is repeated, the animal's response diminishes. Instead of running, the kitten stays stock still, and later only twitches its ears at the stimulus. The animal learns not to react, always provided that the stimulus is not accompanied by unpleasant consequences. It has become *habituated*. This is plainly of advantage, for the animal's life activities would be continually inter-rupted if it were always to react to such stimuli. It learns that they do not result in harm, and so continues with its other activities which are probably biologically more important.

Habituation is essentially something which happens in the nervous system, it is not sensory adaptation or any similar process. But the distinction is not always easy to make, it can sometimes be attempted by varying the stimulus and noting whether the response returns. If it does, then the original effect was habituation, because the sense organ is still equally capable of receiving the stimulus most particularly if the response occurs to weaker stimuli. The original experiments upon which the idea of habituation were based involved the reactions of a snail crawling over a board when a certain weight was dropped from a fixed height onto the same board. With repeated stimulation of this sort the snail's initial response of drawing back into its shell died away, and it showed little response to the weight. But if the height or the weight was then changed, the reaction returned immediately.

It is tempting to think of habituation as a relatively simple process – after all, we habituate to many novel things that come into our lives without making any conscious effort. Hinde was the first to show that habituation can involve an amalgam of processes. He studied the habituation of captive chaffinches to stuffed or model owls. The response of chaffinches to a predator like an owl is to mob it in an apparent attempt to drive it off. This is accompanied by characteristic mobbing calls which attract chaffinches and sometimes other birds as well to the scene. On the first presentation of a stuffed owl, the rate of calling rises to a peak after 2–3 minutes. If the owl is presented again at the same time the next day the rate of calling is now only about half it was before, habituation has occurred. However, the chaffinches now start calling sooner. Chaffinches will also respond to a model owl without eyes, but they will respond much better to this if they are first exposed to a model with eyes. Habituation is therefore not a general run-down of response mechanisms, but includes an increase in some aspects of responsiveness. It is usually more or less stimulus specific, but some degree of generalization occurs. For example, if chaffinches are allowed to habituate to a stuffed owl, some reduction in their rate of calling is detected when they are presented with a stuffed dog. It may also be situation specific; Lorenz found that his greylag geese were habituated to his dogs but attacked them when they turned up in unexpected places.

7.3   Conditioning

Under the heading of 'conditioning' come all those kinds of learning which clearly involve the association of some kind of reward or punishment with the response the animal makes. Pavlov discovered what came to be known as the conditioned reflex in animals. A sound stimulus, like a bell or a tuning fork, has little noticeable effect upon a dog. If the dog is given food immediately after the sound stimulus every time during a period of training, it will eventually salivate directly on hearing the sound, even if no food is given. Salivation has become a conditioned response, evoked by a stimulus that was previously neutral. The food acts as *reinforcement* for the learning process, but if the sound stimulus is subsequently given repeatedly without food being provided the response gradually wanes; a phenomenon known as *extinction*.

Operant or instrumental conditioning is similar in many ways to classical conditioning. It differs mainly in that a voluntary response made by the animal gives rise to reinforcement. Initially, the animal has no expectation of reward when it makes the action. This type of learning has been studied extensively in the laboratory by the American psychologist, B. F. Skinner, who developed a special apparatus known as the Skinner box. This is commonly used with animals such as rats and pigeons. Pigeons, which tend to peck at all kinds of objects, peck sooner or later at a disc in the front panel of the box, which operates a solenoid, and a piece of corn is delivered near the feet of the bird. The pigeon soon 'catches on' and will continue pecking at the disc. Once the response is established, reinforcement need not be given to every peck, but, say, to every ten pecks. Or it may be given at regular time intervals, and can even be of irregular occurrence. Pigeons have been known to continue pecking when reinforced only once every 800 pecks. Learning with partial reinforcement of this kind is very difficult to extinguish – this is very well-known to people who design fruit machines, in which part of the conditioned response is putting money into the machine!

Reinforcement in either classical or operant conditioning need not be a reward, it can be punishment, such as electric shock or some other noxious stimulus. It is now abundantly proved that many insectivorous animals, such as birds, frogs and toads, learn very rapidly to avoid insects that are distasteful because they sting or sequester toxins in their body tissues. Insects which are banded in yellow and black, like some hoverflies, benefit from their resemblance to wasps (Chapters 9). One encounter of a bird with an unpalatable insect such as a monarch butterfly fixes an aversion for monarchs and other butterflies which share the same colour pattern, even though palatable.

Conditioning techniques provide very sensitive methods for the study of sensory perception. Honeybees have been conditioned in the laboratory by exposing them to a scent and then touching their antennae with sugar water. This elicited proboscis extension and feeding, so that eventually a (classically) conditioned response of proboscis extension occurred to the scent alone. Discrimination of that scent from similar ones was then assayed by the occurrence of the conditioned response. Using operant conditioning, the colour vision of pigeons has been explored by training a bird to peck at a disc illuminated with light of a chosen wavelength until a steady response rate was obtained. The wavelength was then gradually changed until the response rate fell sharply, indicating that the bird was beginning to recognize a difference in wavelength. In a study of the ability of electric fish to discriminate between objects of different conductivity (Chapter 2) they were conditioned to swim towards a porous ceramic tube, behind which food was placed (Fig. 7–1). After training, they readily distinguished between a tube containing aquarium water and one containing paraffin wax. By elaborating the training process it was possible to show that the fish could detect a glass rod 2 mm in diameter concealed in a porous pot, but not one of 0.8 mm diameter.

In the study of perception, it is possible to use the reinforcer as the parameter that is varied. Male chaffinches will learn to hop on to a special

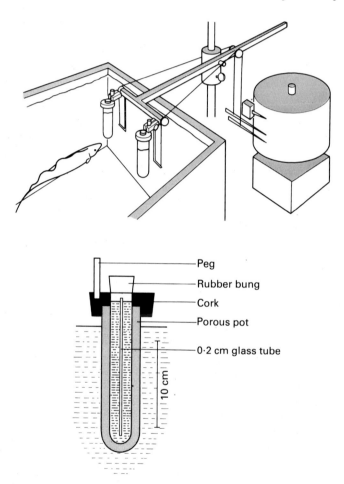

Fig. 7–1 Apparatus used to test the ability of electric fish to discriminate between porous pots containing substances of different conductivity. The fish are conditioned to go to one of the pots by giving them food when they do so, and the number of times they visit each pot is recorded by a simple mechanical system used to mark a revolving drum. (After LISSMANN, H. W. and MACHIN, K. E. (1958). *J. exp. Biol.*, **35**, 451–86.)

perch in order to switch off a recording of mobbing calls. They can also be conditioned to produce the same kind of response in order to be able to hear the song of another chaffinch. It is perhaps surprising, initially, to realize that calls and song can act as reinforcement, but visual stimuli can do so as well. Siamese fighting fish can be trained to swim through a hoop for the 'reward' of being able to display to themselves in a mirror; male zebra finches find a look at their mates sufficient reward for a perch-hopping response, and rhesus monkeys kept in a dimly-lit box will repeatedly press a lever which opens a window through which they can see another monkey, or even just look into an empty room.

To psychologists such as Skinner, it is of great interest that new behaviour patterns can be formed by operant conditioning. This is the type of learning process that has been exploited by animal trainers for centuries. Circus animals have been trained to sit up and beg, jump through fiery hoops, and so forth. More recently, pigeons have been taught to play table tennis and ducks to play the piano with their beaks, dolphins to perform aerobatics and, so it is said, to place limpet mines on ships' bottoms. There is little doubt, also, that a great many human response patterns are built up through similar conditioning processes in which the reinforcement may be a nebulous factor such as the approval of society at large or improvement of self-esteem. Attributes like these, however, are difficult or impossible to measure and so skinnerian theory has limited explanatory value in such complex situations.

7.4   Latent learning

Behaviour in which obvious and immediate reinforcement appears to be lacking has been called latent learning. Because the reinforcement is not obvious to us it does not mean that it is not there or that latent learning is any different from conditioning and so the term is one of convenience, and not one to describe a separate behavioural mechanism. Animals usually learn to recognize features of their environment very rapidly in establishing their territories. This information is subsequently of great value to animals such as mice, and birds, for avoiding predators, obtaining food, and locating their sexual partners, but seems to be obtained in the first instance 'for the experience'. Solitary wasps learn the features of their surroundings in order to identify their nest holes. The bee wolf wasp, *Philanthus triangulum*, for example, digs a hole to which she returns with honeybees whose paralysed bodies she leaves in the hole as a food store for the larva which will hatch from the egg she deposits on them. On leaving the hole to hunt another bee, she may make an orientation flight of some thirty seconds duration. She does this particularly when there has been disturbance of the surroundings of the nest. If a ring of objects is put around the nest entrance before she leaves, and moved sideways while she is away, on her return she searches for the nest-hole in the middle of the ring in its new position. In addition, these experiments can indicate how the perceptual world of the insect is organized. It is plain that the wasp recognizes configurations, like circles, rather than the objects which make them up. 'Circleness' is recognized even when the number of objects making up the circle is reduced from sixteen to eight (keeping the diameter the same). Other wasps learn the more distant surroundings of the nest area as if they were arranged in a series of circles centred on the nest hole, but these objects are learned at other times than during the orientation flight. This enables the insect to learn in a very short time and the memory may remain for a matter of days.

7.5   Imprinting

It is often found that there are sensitive periods in the life of an individual during which certain things may be learnt and become thereafter relatively fixed and resistant to change. Song learning in birds is one example: the sensitive period for this comes near the end of the first year in chaffinches, and after the thirteenth month no further learning appears to take place. Depriving young animals such as dogs or rhesus monkeys with companions during the first few months of life has lasting effects on their behaviour, so that they may show antisocial behaviour towards members of the same species in adult life. There is a well-marked sensitive period in young chicks immediately after hatching. Like goslings and ducklings they learn the visual appearance and typical calls of their parents whom they will then follow. This process was called *imprinting* by Lorenz. If goslings are not allowed to see their parent, but exposed instead to a human being, or a moving box, then they may follow that instead. Ducklings have been imprinted upon matchboxes, and moorhens on a large canvas hide that was moved around.

It was once thought that imprinting was a special form of learning confined to the first $1\frac{1}{2}$ days of the animal's life, and irreversible. It is now understood as a process in which the animal learns to recognize features of its immediate environment, so that it is then able to discriminate any new or unusual feature, from which it may withdraw. Chicks reared in groups, for example, will imprint on one another and so will not follow an unfamiliar figure. However, a chick reared in isolation has its sensitive period extended. It is important to note that the following response is not imprinting, but only a means by which it can be measured. A chick will usually give pleasure calls when it is united with an imprinted object, and distress calls when it is separated from it. These calls can also be used to assay imprinting.

It was earlier believed that imprinting would determine the choice of mate for an animal when adult. This is at least partially true, for there are many accounts of hand-reared birds courting human beings. Some ducks reared by foster parents of different species will attempt to court mates of the foster species, often ignoring potential mates of their own kind, though others show a preference for their own species. Experience during adolescence is capable of changing a sexual preference that has arisen through early imprinting.

7.6   Insight learning

There are some learning processes in which an animal produces a new kind of response through insight, a phenomenon which depends very much upon perceptual abilities. The use of tools is sometimes, but not always, an example of insight learning. If chimpanzees are shown bananas just beyond their reach they are often capable of stacking boxes and climbing on them, or fitting sticks together in order to reach the food. Such problem solving appears to depend upon the animal realizing in some way what needs to be done, and so trial and error alone cannot explain the behaviour, but concept formation must also be involved. However, without knowing the previous history of the animal it is difficult to be sure that such a response is a new one. In the field, chimpanzees have been observed to use sticks to fish for termites through openings in mound nests. They have been seen also to use sticks as missiles in attempts to drive off a stuffed leopard. Responses such as these can be spread by imitation. A Japanese macaque monkey which developed a habit of washing sweet potatoes before eating was imitated by her companions and a year later this was a normal activity for half the troop.

Rather better evidence for animals solving problems by grasping a simple concept comes from studies in the laboratory. An animal is given two different objects, movement of one of which reveals a piece of food. The presentation is then repeated several times. After this test has been repeated about 100 times, using different objects, rhesus monkeys begin to make the correct choice on the second presentation in each test. They gradually improve, and after about 300 trials they nearly always choose correctly the second time. They have learnt the principle involved. Marmosets learn this much more slowly, and rats and cats more slowly still. A few attempts have been made to teach chimpanzees to speak, but only a few words have ever been obtained from them. More success has been obtained by teaching chimpanzees sign language, in which the animals were not only able to name a wide range of objects but also showed that they understood concepts like 'nice', 'dirty', 'green' and so forth.

8 Social Behaviour

Few animals live solitary lives, most needing at least to find a mate from among their own species. Many animals live in groups which may be temporary or permanent. Frequently, grouping together requires some behavioural adaptation which ensures the cohesion of the group. This may require a complex of special behaviour patterns like those found in truly social animals, but the less permanent aggregations of animals involve simpler processes.

At one time it seemed that a distinction could be made between an aggregation – a group of animals which come together for some external reason – and a society where the parents and offspring remained together for long periods kept in association by mutual behaviour. It seemed that aggregation did not involve mutual behaviour, but this distinction seems more and more difficult to sustain. Sometimes the animals do not seem to react to each other as animals. Thus, brittle stars in a tank containing no stones or sand will cluster together intertwining their arms. This seems to have the sole effect of giving them maximum contact, for they will equally well entwine with glass rods. So here there does not seem to be mutual behaviour, the brittle star's 'companion' can be an inanimate one or one of its own kind. On the sea bottom, however, brittle stars aggregate only with one another and continue walking past other objects until they find conspecifics. Collections of woodlice are found under bark, no doubt grouped in such places as a result of each one's individual responses to humidity, light and contact. But the group is now known to be held together by pheromones as well. Thus, like the brittle stars, they are responding to stimuli that only other individuals can provide.

Aggregations consist of animals of the same species grouped together in the same place, each one acting essentially as an individual and not co-operating with others. In a society, individuals co-operate, and the grouping can be seen to have an organization of its own. A society may arise as a family group or as a number of adult individuals which come together and co-operate. Within a society, individuals tend to specialize in what they do, which results in a division of labour. There is a complex system of communication, which involves mechanisms for recognizing other members of the society and strengthening the bonds between them, and so allows discrimination against outsiders.

In recent years, the importance of altruism in the evolution of social behaviour has been realized. The simplest division of labour found is into those individuals that are engaged in breeding and those which are not. The latter will work for the reproductives and young in the group and may even readily sacrifice themselves in defence of the society. Natural selection cannot perpetuate these altruistic traits by acting on these animals as individuals, because they will not have their own offspring. Selection must act on the reproductives favouring those genes which are conducive to altruistic behaviour in members of a family. In social Hymenoptera (bees, wasps, and ants) there is a great similarity in the gene complement of the queen and the workers, which are her sterile daughters developed from diploid eggs, but an even greater one between the sister workers. It is not surprising, therefore, that workers devote themselves mainly to rearing other workers, because they are more likely to pass on their own genes by contributing to the potential and success of the colony in this way than by having children themselves. The evolution of altruism cannot be so elegantly explained in termites, which do not have the haploid-diploid system of caste determination.

8.1   Insect societies

Most of the societies of bees, wasps, ants, and termites consist of one or a few reproductive females with large numbers of their offspring. In the more complex of these societies, there is usually only one queen. After mating, the male remains with the new queen in termites, but not in the Hymenoptera. Termite societies also differ in having immature forms which function as workers, while the immature individuals of hymenopteran societies are legless larvae. Some of the more primitive societies contain a number of queens. The *Polistes* paper wasps provide examples, in which a new nest is often founded by a group of females, one of which becomes dominant to the others. She spends most of her time on the nest, her ovaries develop fully and she lays many eggs, eating any laid by other queens. This system, dependent upon a dominance hierarchy, contrasts with that in honeybees in which two queens in the same hive fight vigorously until one is killed. The reproductive dominance of the remaining queen is then maintained by the 'queen substance' pheromone (Chapter 6).

Integration of insect societies depends very much upon food exchange and mutual licking and grooming (*trophallaxis*). Trophallaxis serves to distribute food, odours and pheromones throughout the colony. Foragers returning to the hive are solicited by other workers and transfer regurgitated food to them. A worker will solicit food from isolated heads of other worker bees, and even from plasticine pieces of the right size, if they have two 'antennae' of wire fixed in them.

All the members of a colony share the same food supply through the trophallactic exchange stream which is circulating round the colony. Radioactive tracer-marked sugar is passed to over half of the members of a honeybee hive within 24 hours and to nearly all of them within five days. There is little doubt that this mutual sharing of food is one source of the common colony odour borne by all members of the society. This odour acts as a badge which distinguishes member from non-member, i.e.

friend from enemy of another colony or species. As an ant, for example, needs to be isolated from the colony for only a matter of hours before it evokes at least the beginnings of aggressive behaviour when it returns to its colony, it is plain that the odour is changing with time, no doubt because one food source becomes exhausted and the foragers move on to another.

At the entrance of a bee hive, a number of bees can be seen facing away from the hive entrance and poised, for most of the time, on the tips of their legs. They react to small passing shadows and appear obviously alert. These are the guard bees who challenge every insect landing on the alighting board of the hive. Those with the correct odour are permitted to enter the hive. Those with the wrong odour are held by a leg and inspected. If they are bees which have wandered in from a neighbouring hive, they take up a submissive posture as the guards inspect them. They may strop their tongues, a movement that occurs in food transference. Finally, they are pushed off the alighting board unharmed. If the stranger is a bee bent on robbing the honey stores, she will fight back when handled until finally the guards will sting her to death. The behaviour of the guards depends upon the odour and behaviour of the bees which land on the alighting board.

The efficiency of the colony as a means of raising large numbers of young depends upon a regular food-supply. Honeybees fly to their food sources in flowers which they learn to recognize from land-marks in the vicinity, just as they learn to recognize the hive's position on their return. They utilize their time sense to restrict their visits to particular groups of flowers at the time of day when their nectar is flowing. It is clearly more efficient if a successful forager can recruit others to come to a plentiful food supply. Honeybees can direct other foragers to sources they have found. This they do by performing on the vertical face of the comb within the hive a dance, which is followed by a few other bees. Below about eighty metres distance, they do not indicate anything of position of food but only stimulate other bees to forage within that distance for food distinguished by the scents carried on the dancer's body and by the taste present in fluid she regurgitates. But after returning from food sources farther from the hive, a forager performs a waggle dance by which she indicates distance and direction of the source (Fig. 8–1). The dance is in the form of a figure of eight; the speed of dancing and therefore the number of waggles of the dance per minute indicates the distance while the direction of the straight part of the dance shows the direction of the food with respect to the sun's direction.

The bee gets its direction information from the sun either directly or, if the sun is overcast, from the pattern of polarized light in the sky, and uses this to orientate its dance correctly. Needless to say, allowance has to be made for the sun's movement and a 'clock' must be involved. An instance of prolonged dancing in the hive long after sunset showed that the forager was indicating the direction of the food it had found before dusk with

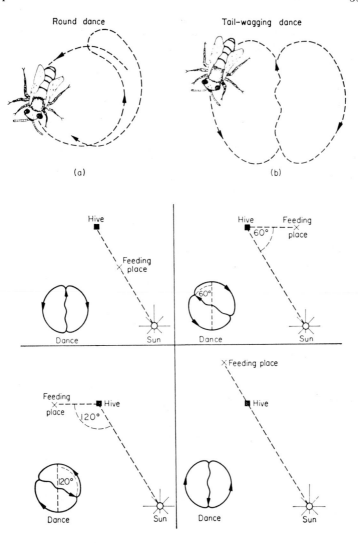

Fig. 8–1   The dances of honeybees. (a) The round dance performed when food is less than 90 metres from the hive. (b) The waggle dance and, below, the relation between the angle of the dance and the vertical with the angle between sun and food. (After FRISCH, K. VON (1954). *The Dancing Bees*. Methuen and Co. Ltd., London.)

respect to a non-existent sun, dancing as if the sun had continued across the visible sky instead of sinking below the horizon.

Since the dance is evoked only from a forager who has found a good food source, the attention of foragers will be kept on the plentiful food supplies. As the food decreases the vigour of the dancing decreases and finally foragers from the reduced source no longer dance.

Dialects are found in this bee 'language'. The Italian race of the honeybee has a third dance, called the sickle dance, which is used for intermediate distances. Dance tempo also varies among races, so that a given waggle dance will indicate distances that are different for each race of the European honeybee.

Recent evidence indicates that honeybees do not always rely upon their dance language for finding their way to food. Experienced foragers seem to acquire a very detailed knowledge of their foraging range and will go to certain flowers without further direction if only the scent of these is blown into the hive. Bumble-bees also learn foraging routes, which they gradually modify to take in new nectar sources as they relinquish declining ones. Species of *Trigona*, South American stingless bees, mark the way to a food source with scent spots which are put on to foliage or stones at intervals. A successful forager recruits others by leading them along the trail to the food. Ants also use scent trails of various kinds (Chapter 6), but some species have a primitive form of recruitment known as *tandem running*, in which a successful forager solicits another worker to follow her in tandem to the food, stimulated by an attractant chemical secreted on to the tip of the sting.

8.2 Vertebrate social behaviour

Insect societies are fountains of complex behaviour which are often very stereotyped. We must be careful, however, not to over emphasize this: honeybee foragers, for example, can be conditioned very rapidly to respond to colours, scents and patterns, and some ants learn to recognize their own brood. Experience may play a far greater part in the integration of insect societies than is commonly conceived at present, but in vertebrate societies it is certainly the factor of paramount importance. This comes into the relationship between mates, as the establishment of the pair bond, into the mother-infant relationship, and into the relationship between adults. The latter finds expression in a dominance hierarchy, typical of vertebrate societies. In its simplest form this is the linear peck-order found in birds such as pigeons and farmyard hens. In hens, there is one dominant α-individual who can peck any other with impunity. Below her, there is a β-individual who can peck any other except the α-bird, and so on through the line, each being able to peck the hens below her but not those above. The hierarchy is not necessarily linear among birds, but may be organized in triangles, so that A pecks B, who pecks C, who pecks A, or in more complex figures. The α-individual has certain rights because of its rank: often this means the first choice of the best roosting places and of the most attractive food. Males often have the first choice of the available females.

In birds, dominance is based upon experience and individual

recognition. Once it is established in a group it may remain fairly stable, perhaps for the lifetime of some individuals. However, the peck order varies with species and with conditions. Illness may reduce an individual's vitality and make it liable to lose its place to a rival. Injections of male hormone can raise an individual's rank in a hierarchy, increasing its general aggressivity. In domestic hens, at least, more subtle factors also operate, such as whether an individual is in moult, physical resemblance to previous high-ranking individuals, and so forth.

On the whole, a hierarchical system decreases the amount of inter-individual aggression which occurs in a group, because the subordinate animals respond to signals of the dominants and give way to them without contest. This is advantageous because fighting not only causes injury but takes up time which is wasted from the point of view of the group. In some vertebrate societies, however, overt aggression occurs when young males are forced to leave the group when they are almost fully-grown. There is no place for them so long as the a-male, usually their father, is active. They have to seek mates elsewhere, or fight him. Contests may be very bloody on these occasions; those of elephant seals are a good example. The dominant bull rears up and crashes down upon its rival, attempting to injure him with his teeth. Many calves are injured or killed quite fortuitously in these combats by the weight of the fighters coming down on them. The old bulls are covered with battle scars of their contests.

Such fights in which blood is drawn are unusual. More typical are the struggles of rival red deer contesting the leadership of the herd. With horns locked, the two males lunge and strain. In this way they seem to get the measure of each other's strength and the weaker one fairly soon seeks a way of breaking off the fight and running away.

8.3   Primate societies

Division of labour in primate societies in which individuals tend to adopt and specialize in different roles – defence, food-finding, repro-duction, etc. – is generally most marked among ground-living species such as baboons, which move about in open savanna where they must defend themselves against many predators. The strongest and largest male of a troop of a savanna baboon species is usually dominant, and has, like other adult males, a thick pelt of fur on his head and shoulders which accentuates his size. His position is not determined solely by his fighting ability, but also by his relationship with other males. Indeed, two or three high-ranking males may collaborate and jointly dominate the troop, backing up one another when necessary. Dominant males have precedence, and others move out of the way when they approach: dominants also rapidly intervene to prevent any fighting within the troop and so protect weaker individuals. When the troop is on the move, they stay in the centre with the young juveniles and the females carrying infants. Less dominant males are found at the front and rear, so that the females and young are well protected. If the troop is attacked, the adult males come forward and interpose themselves between the attackers and the rest of the group.

Signals of dominance and submission are very varied in primates and include vocalizations, special postures and facial expressions. They are not always easy to interpret; for example the cheerful-looking grin of the chimpanzee, often exploited in the advertising trade, exists in several subtly different forms, some of them signifying submission or appeasement. A steady stare is a threat in many primates, which in baboons and some monkeys is accentuated by lifting the eyebrows to expose white eyelids. Subordinate individuals may turn their heads away in response. Making faces at captive primates can be very instructive, providing you know what your expressions are likely to mean to them.

Many forest monkeys express threat by shaking leaves, and gorillas threaten by beating their own chests. Ring-tailed lemurs have 'stink fights' using scent as a threat. In such contests, a lemur draws its long tail against scent glands on the forearms and chest, and then holds it arched over the back, quivering violently so that the odour appears to be shaken forward towards the opponent.

A dominant primate generally has a 'confident' relaxed manner of walking (Fig. 8–2) and approaches subordinates directly. A subordinate turns and looks aside, or moves out of the way. Its locomotion is more hesitant and its posture hunched. Presenting of the genital area is also a common subordinate signal, to which the dominant individual responds by grasping the haunches of the other animal or actually mounting it.

The dominance hierarchy of primate societies works well and ensures peaceable co-operation in the group providing the individuals know their status and have learnt how to respond to all the other members of the group. This learning process begins in the infants with the mother-infant bond. Much of what we know about bonds in primate societies has come from the work of the Harlows on captive rhesus monkeys. Initially, the attachment of the infant to its mother is very close and the infant does not stray far, returning to cling to her if threatened in any way. Infants separated from their mothers and reared in isolation will cling to a piece of towelling on a wire frame provided as a substitute. Without even this during the first eight months, infants are unusually fearful of novel objects they encounter. Monkeys reared on wire and cloth 'mothers' became very poor mothers themselves, often rejecting their infants or becoming aggressive towards them. They also showed very abnormal sexual behaviour.

As the infants grow older, they spend less time close to the mother who in fact begins to reject them more and more as time passes. They then interact more and more with other individuals, particularly with other young juveniles. During play with others they appear not only to develop their motor skills but to gain a familiarity with others through contact and

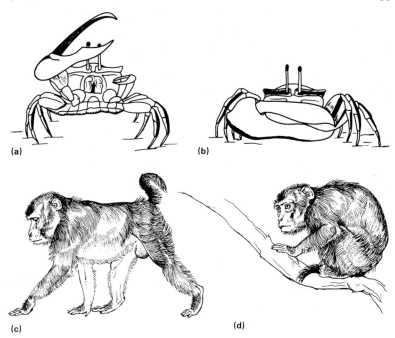

Fig. 8–2 Postures of threat and submission often share common features in a wide range of animals (see also Fig. 4–4). (a) High intensity threat of a fiddler crab and (b) submission and appeasement. (c) Dominant male rhesus monkey: note the straight back and tail and (d) cringing posture of subordinate female rhesus. ((a–b) After CRANE, J. (1966). *Phil. Trans. Roy. Soc. (B)*, **251**, 459–72; (c–d) redrawn by Priscilla Barrett after HINDE, R. A. (1966). *Phil. Trans. Roy. Soc. (B)*, **251**, 285–94.)

contest that persists into their adult life. In this way, bonds develop between the offspring of different females, which gives the society cohesiveness. This is further aided by young infants being generally very attractive to adult males and females and forming foci of attention.

The bases of the adult behaviour involved in pair formation and co-operation among adult males are thus laid down during infancy and adolescence. Even the correct postures for mating are learnt in rhesus monkeys during play among adolescents. Mutual grooming is important in forging and maintaining these bonds, although not in the same way as among social insects. Grooming appears to be pleasurable to primates, and allows them to come into close contact. It is related to status, for example in rhesus monkeys dominants usually receive grooming, but rarely groom subordinates. It also reduces tension between individuals and is sometimes used by a mother to calm her excited infant. In general, frequent contact seems to strengthen bonds. In the dusky titi, a small South American monkey, mated pairs frequently intertwine their long tails in a sort of plait.

The outcome of the complex behavioural interactions of a typical primate society is peaceable stability. But this does not mean that the dominance hierarchy is immutable, on the contrary, it is a dynamic system that changes as individuals in the group mature, form new pair bonds, and have babies. In macaques, females attain a higher status if they pair with a high-ranking male. The young of high-ranking females also tend to achieve high status, partly because they mix more with the young of similar rank to their own mothers. The importance of experience in establishing the social structure of a group explains why fighting and bloodshed is common in artificially constructed groups of primates in zoos: here, despotism can arise unchecked and animals act as individuals, failing to co-operate with one another.

9 Behaviour and Survival

Much of this book has been about the ways in which animals react with others of the same species. Courtship and social behaviour patterns often involve conspicuous signals and striking behaviour sequences, but animals spend most of their time making themselves inconspicuous or defending themselves against predators often while trying to feed at the same time. Even periods of inactivity, like sleep, can be regarded as adaptive behaviour because sleeping animals can remain concealed.

9.1 Crypsis

Animals which show a visual resemblance to some part of their environment are said to show *crypsis*. Crypsis is usually associated with postures and resting positions that match the animal even more closely with its background. Mantids and stick insects that mimic grass or twigs align themselves with the vegetation and press their limbs close to the body. In some, the forelegs are extended forwards and so conceal the contour of the head. Cuttlefish, which can change their colour pattern to match different substrates, disturb the sand or gravel where they settle so that it covers the edges of their fins. The young larvae of some swallowtail butterflies closely resemble bird droppings, and rest with the body flexed, which enhances the mimicry.

9.2 Displays of cryptic animals

When cryptic animals such as lizards, rabbits and snipe are disturbed they suddenly explode into action, which tends to confuse us, and perhaps all predators. Such animals often show sudden and seemingly random alterations in course, which make them very difficult to follow. Such unpredictable behaviour is called 'protean' behaviour. Some grasshoppers and moths have brightly coloured underwings – known as flash colouration patterns – which make them conspicuous in mid-flight, but they often deviate from the advertised trajectory when they land and conceal their hindwings once more.

Many animals achieve a startling effect by suddenly and dramatically changing their form and appearance without immediately running or flying away. This is known as deimatic behaviour. Many cryptic mantids have particularly well-developed deimatic displays (Fig. 9–1b). *Mantis religiosa* rears up on its legs, spreads its wings to the full and holds its raptorial forelegs high in the air splayed wide. This is accompanied by a swishing sound made by the abdomen moving over the hind wings.

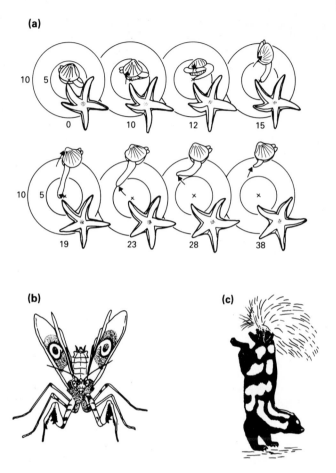

(a)

(b)

(c)

Fig. 9–1 Examples of anti-predator behaviour. (a) Drawings from a filmed sequence of a single leaping movement in a cockle reacting to the approach of a starfish. The position of the tip of the foot which presses on the sand is marked with a cross. (b) Deimatic display of the African target mantis. (c) Deimatic display of the spotted skunk. ((a) After ANSELL, A. D. (1967). *Anim. Behav.*, **15**, 421–6; (b) after MACKINNON, J. (1970). *Z. Tierpsychol.*, **27**, 150–5; and (c) after BOULIÈRE, F. (1955). *The Natural History of Mammals.* Harrap, London.)

Somewhat similar behaviour is seen in many birds, as Darwin observed; owls, hawks and young cuckoos, among others, extend their wings and tail plumage when disturbed, opening the mouth wide (in cuckoos) and hissing and clicking with the beak (in barn owls). The cobras have a deimatic display which involves inflation of the hood, and the skunk *Spilogale* erects its tail and advances towards a predator on its forelegs, with its body held vertically (Fig. 9–1c).

Another kind of deimatic behaviour involves the sudden presentation of eyespots. In some insects, like the eyed hawk moth and many of the silk moths (Saturniidae) there are large eyespots on the hind wings, which may be exposed suddenly, accompanied by rocking movements of the body.

A. D. Blest showed that two large artificial eyespots deterred insectivorous birds from feeding on mealworms placed between them. They present many features of a predatory animal – large eyes, both facing forwards for good binocular vision – and in some cases are thought to startle some animals long enough to allow the insect to make a subsequent rapid escape. In the peacock butterfly there is a large eyespot on the upper surface of each wing, but the wings are normally folded at rest, showing only the cryptically coloured undersides. When disturbed, the butterfly opens and closes its wings repeatedly, and makes a clicking sound. In Denmark, it hibernates in caves with roosting bats, and ultrasonic components of the clicks act as a deterrent to the bats which use ultrasound in their sonar navigation.

9.3   Warning colouration

Warning colouration is a loose and subjective term: we can be more objective by talking about *aposematism*, and *Batesian* and *Müllerian* mimicry. An *aposematic* animal may be poisonous or have a well-developed weapon like a sting, and advertises this to predators by bright colours, characteristic odours or sounds it produces. Let us consider a wasp, which cannot only deliver a painful sting if molested, but also recruit other wasps, it is strikingly coloured and can produce an angry buzz. Many species of hoverfly, and other insects also, have the same colour pattern and can also make buzzing noises, but are quite harmless. They are said to show *Batesian* mimicry, after the naturalist H. W. Bates who first noticed the phenomenon among Brazilian butterflies in the nineteenth century. Batesian mimicry has been well studied in some butterfly families, including the swallowtails and danaids (monarchs and viceroys). The monarch butterfly, *Danaus plexippus*, feeds as a caterpillar on milkweeds which contain poisonous cardiac glycosides. The larva stores the toxins and they remain in the butterfly. L. P. Brower showed that when a naïve jay was given a monarch as food it became immediately violently sick as a result of the toxins and subsequently rejected not only other monarchs but the palatable mimetic species *Limentis archippus*. Similarly, toads learn very quickly to reject honeybees and thereafter also reject dronefly mimics of the bees. The success of Batesian mimicry therefore depends very much upon the rapid conditioning of predators.

It has been shown that birds which have learnt to avoid the brightly-banded yellow and black caterpillars of the cinnabar moth, which is poisonous, then avoid wasps. This is an example of generalization on the part of the predators (see Chapter 7). Unpalatable insects that share the same colour patterns, even if they are not excellent mimics of one another, will gain an advantage because the colour patterns and the message they carry will be more widely advertised. This is *Müllerian* mimicry (Müller was a contemporary of Bates). Many wasps have yellow and black banding, or in the tropics, dark blue colouration of the body or a bluish colouration of the wings. Such colouration is often shared by other sting-bearing Hymenoptera (e.g. bumblebees, solitary bees) that live in the same habitat. Many New-World snakes are very conspicuous, banded in red, black and yellow, and are called coral snakes. They include both Batesian and Müllerian mimics, and also some that are so deadly that no attacker would live to learn from its experience. They must, therefore, mimic snakes that are less deadly and give a predator a chance to learn from its encounters. They are known as *Mertensian* mimics.

Müllerian mimics are usually not only similar in colouration but in behaviour: in insects this is sometimes shown by relatively slow flight patterns with none of the strategies of concealment at rest and evasive manoeuvres shown by others. Butterflies of the family Heliconiidae, which are very common in Latin America, sometimes occur in loose swarms in which a number of heliconiid species and also Müllerian mimics from other families take to the air when disturbed and fly together and rest in mixed assemblages, conspicuous on the foliage.

9.4   Chemical defences

We have seen that the success of aposematic animals depends upon their ability to harbour toxins in their bodies. Many invertebrates (as well as some vertebrates, such as skunks) have exocrine glands which ooze or squirt chemicals used in defence. The harvestman, *Vonones sayi*, regurgitates a watery fluid which accumulates at the edge of the body. There, glands discharge repellent chemicals (quinones) into the droplets of fluid. The animal then dips the tips of its forelegs into the secretion and spreads it on to the attacker. The bombardier beetle, *Brachinus*, manufactures quinones in an explosive chemical reaction: in special glands at the tip of the abdomen it mixes hydroquinones, hydrogen peroxide and catalysts, with the result that any animal behind it is blasted with a mixture of quinones and water vapour at 100°C.

Whip scorpions and wood ants, *Formica* spp., are among those animals that are able to spray a jet of acid from the tip of the abdomen at other ants or at predatory birds. The soldiers of some termites are highly specialized for chemical warfare: they are called 'nasutes' and have the front of the head capsule prolonged into a nozzle. A jet of liquid is squirted through this which rapidly becomes very viscous on exposure to air. It sticks to the cuticle of other arthropods and entangles them; in some species it carries toxins which will kill quite large ants.

Some water beetles have a well-developed defensive chemical arsenal. In the whirligig beetle, *Gyrinus natator*, this extends to a secretion which is spread over the surface of the body by grooming and which is toxic to a wide variety of micro-organisms living on the water surface which might otherwise attack the cuticle of the insect. A beetle, *Stenus*, that hunts on river banks, has a secretion that it uses as a propellant in rapidly regaining the bank when it accidently falls on the water surface. This is produced by glands at the tip of the abdomen and acts by lowering the surface tension of the water, causing the insect to shoot forwards at 40–50 cm s^{-1}. Water skaters, *Velia*, emit a similar propellant from the proboscis.

9.5   Detection of predators

The American biologist, T. C. Schneirla, emphasized that animals respond evasively to changes in the steady state of stimuli in their normal environment: to sudden and rapid increases in visual stimulation, for example, or to harsh irregular sounds. This will alert them to predators, but many prey animals have special adaptations for detecting predators. Rabbits, rodents, and birds such as pigeons have their eyes on the sides of the head, which gives them good all-round vision. In most predatory vertebrates, in contrast, the eyes face forwards, and good stereoscopic vision is achieved at the expense of panoramic vision.

Detection of predators by chemical means is well-developed in many aquatic molluscs. Scallops are highly sensitive to chemicals on the tube feet of starfish, in response to which they swim away by violently clapping their shell values together. Cockles escape by lashing movements of the extensible foot (Fig. 9–1), and a similar response is shown by gastropods such as the mud snail, *Nassarius*. Similar behaviour is shown by freshwater gastropods in response to contact with leeches.

The work of K. D. Roeder and his co-workers in the U.S.A. has shown how certain families of moths are able to detect insectivorous bats in flight and take evasive action. Moths of the superfamily Noctuoidea have a pair of ears (tympanal organs) on the first abdominal segment, each containing only two sense cells attached to the tympanum. In response to a distant bat producing ultrasonic echo-location calls, one of the two sense cells is activated. The tympanal organs are directionally sensitive, and the moth turns and flies away from the source of the sound. If this is ineffective, the second sense cell in the tympanum is stimulated when the moth is within range of the bat's sonar (about eight metres). Depending on the species, the moth then takes rapid evasive action, falling, diving, or performing a sudden aerial loop. The hawkmoths, which are rapid and very powerful fliers that often feed while hovering, have no tympanal organs but a segment of each palp which is greatly swollen and acts as an ultrasound detector. In response to the cries of bats, hawkmoths fly off at high speed: it seems that they can fly sufficiently rapidly to escape without sudden aerial acrobatics.

9.6   Tactics of predators

There are many predatory animals that do not hunt their prey but ensnare them in various ways. A variety of different techniques is used by spiders. The orb-web spiders spin a sticky thread on a taut framework of threads. They detect snared prey by vibrations set up in the threads, which are sensed by the lyriform organs at the tips of the legs. Spiders of other species use silk threads in quite different ways. The trap-door spider lives in a burrow capped with a lid of silk and particles of earth. It leaps out of this on to prey walking near the trapdoor. The Australian bola spider hangs from a horizontal thread and swings from one leg a thread with a sticky blob on the end which is attractive to some moths. The New Zealand 'glow-worm', the larva of the fly *Arachnocampa*, lives in caves suspended by a horizontal web. Numerous sticky threads hang down from this web and trap insects attracted by light emitted from the luminous abdomen of the larva. Luminous lures are also found in many deep-sea fish, often as appendages in and around the mouth.

Some spiders throw a thread with a sticky blob on the end at passing insects. Somewhat similar behaviour is seen in the larva of the ant-lion (a neuropteran). This makes a small pit in sandy soil into which ants are liable to fall. If sand grains fall to the bottom of the pit where the insect lies buried, it flicks sand upwards which usually helps to dislodge the prey so that it falls right to the bottom of the pit where it is seized. The archer fish, *Toxotes*, traps insects by spitting a jet of water droplets at them from the water surface and can hit them from a distance of over a metre.

Some animals have evolved specialized non-visual sensory mechanisms for detection of cryptic prey or of prey in the dark, we have already considered bats in this respect. Dogfish, skates and rays are able to detect cryptically coloured flatfish, even when they are buried below the surface of sand. They do this with the aid of electroreceptors in the skin which pick up muscle potentials controlling the respiratory movements of the flatfish. Some snakes, including rattlesnakes, have pit organs below the eyes which are extremely sensitive to radiant heat, detecting temperature differences as low as $0.002°C$. A rattlesnake can detect a live mouse 15 cm distant and strike at it accurately without being able to see it. A blind snake, *Leptotyphlops*, feeds on nomadic army ants and is able to follow their trails because it can detect their trail pheromone. Some predators are able to prey heavily on cryptic animals, where they are common, by forming a search image.

There are a few predators that benefit from mimicry. One example is a firefly, *Photurus*, which replicates the flashing code of *Photinus* females, and feeds on the males of the latter species as they land. Another example is the false cleaner fish of coral reefs which mimics fish that clean the body surface of parrot fish. Both the cleaner and false cleaner have an undulating swimming pattern which induces the parrot fish to stay still.

The cleaner then performs its service while the false cleaner bites a piece off the parrot fish.

Cryptic behaviour, in which predators remain more or less concealed until they are able to strike at their targets, is very widespread, and occurs in animals such as mantids, chameleons, cats and lions. The domestic cat and the lion both stalk prey in the same way, running quickly towards it with the body flattened close to the ground, and then remaining still, poised for a final sprint and spring. Leopards hunt similarly, or by waiting in ambush, while tigers were reported to have been conditioned by the sounds of gunfire during the Vietnam war, and went towards this to take the casualties. In contrast with lion and leopard, other predators in the Serengeti region of East Africa have different strategies. Cheetah, wild dog and hyenas capture their prey by running them down rather than by stalking. A balance of numbers of predators and prey is achieved in this habitat because the prey (wildebeest, gazelle, etc.) have a number of effective defences; besides out-running the predators, they can attack with hooves and horns, which they sometimes do in a group with the young and pregnant females protected within the herd.

The use of tools for feeding is very rare in animals, and where it occurs it is often difficult to decide whether it depends on imitative abilities. It almost certainly does not in the woodpecker finch of the Galapagos, which uses a cactus spine for prizing insects out of the bark of trees. In a somewhat similar way, chimpanzees use long twigs which they push into holes in termite mounds. Soldier termites grasp the twigs and are then eaten by the chimpanzee. Some animals have developed methods of

Fig. 9–2 An example of specialized predator tactics. Ithomiine butterflies are common Müllerian mimics in S. America and usually escape predation. In some regions, tanagers have learnt to feed on these butterflies by squeezing out the nectar-rich abdominal contents and rejecting the rest of the insect. (After BROWN, K. S. and NETO, J. V. (1976). *Biotropica*, **8**, 136–41.)

cracking open eggs or tough-shelled invertebrates such as molluscs. The California sea otter floats on its back holding a stone on its chest and cracks open molluscs by hitting them against the stone. The Egyptian vulture opens ostrich eggs by lifting stones in its beak and throwing them against the eggs. The banded mongoose has also been observed to break ostrich eggs by throwing stones at them, but opens smaller eggs and cracks open millipedes by flinging them backwards between its hind legs at large rocks.

In these examples it is very difficult to determine how the behaviour has arisen, bringing us back to the central issue of how behaviour can be modified and perpetuated. Learning is important in behaviour like that shown in Fig. 9–2 and the development of new 'tricks' can clearly upset an ecological balance between predator and prey, making it necessary for the prey to evolve counter-measures. The behaviour of predator and prey must be subject to continual co-evolutionary pressures, but this is something about which we know very little: without a fossil record of behaviour evolutionary trends can be studied only in the very short term.

Further Reading

ALCOCK, J. (1975). *Animal Behavior, an Evolutionary Approach.* Sinauer Associates, Sunderland, Massachusetts.

BASTOCK, M. (1967). *Courtship: a zoological study.* Heinemann Educational Books, London.

EDMUNDS, M. (1974). *Defence in Animals.* Longman, Harlow, Essex.

EIBL-EIBESFELDT, I. (1970). *Ethology, the Biology of Behavior.* Holt, Rinehart and Winston, New York.

EIBL-EIBESFELDT, I. (1971). *Love and Hate.* Methuen and Co. Ltd., London.

EISNER, T. and WILSON, E. O. (Eds.) (1975). *Animal Behavior* (readings from the *Scientific American*). W. H. Freeman, San Francisco.

HINDE, R. A. (1970). *Animal Behaviour* (Second Edition). McGraw-Hill, Maidenhead.

HINDE, R. A. (Ed.) (1972). *Non-verbal Communication.* Cambridge University Press, London.

MANNING, A. W. A. (1979). *An introduction to Animal Behaviour.* (Third Edition). Edward Arnold, London.

MARLER, P. and HAMILTON, W. J. (1966). *Mechanisms of Animal Behavior.* Wiley, New York.

WILSON, E. O. (1975). *Sociobiology.* Belknap Press, Cambridge, Mass.